T0239192

SpringerBriefs in Applied Sciences and Technology

Series Editor

Andreas Öchsner, Griffith School of Engineering, Griffith University, Southport, QLD, Australia

SpringerBriefs present concise summaries of cutting-edge research and practical applications across a wide spectrum of fields. Featuring compact volumes of 50 to 125 pages, the series covers a range of content from professional to academic.

Typical publications can be:

- A timely report of state-of-the art methods
- An introduction to or a manual for the application of mathematical or computer techniques
- A bridge between new research results, as published in journal articles
- A snapshot of a hot or emerging topic
- An in-depth case study
- A presentation of core concepts that students must understand in order to make independent contributions

SpringerBriefs are characterized by fast, global electronic dissemination, standard publishing contracts, standardized manuscript preparation and formatting guidelines, and expedited production schedules.

On the one hand, **SpringerBriefs in Applied Sciences and Technology** are devoted to the publication of fundamentals and applications within the different classical engineering disciplines as well as in interdisciplinary fields that recently emerged between these areas. On the other hand, as the boundary separating fundamental research and applied technology is more and more dissolving, this series is particularly open to trans-disciplinary topics between fundamental science and engineering.

Indexed by EI-Compendex, SCOPUS and Springerlink.

More information about this series at http://www.springer.com/series/8884

N. A. Elmunim · M. Abdullah

Ionospheric Delay
Investigation and Forecasting

N. A. Elmunim
Department of Electrical Engineering
(ECE), College of Engineering
Princess Nourah bint Abdulrahman
University, Riyadh, Saudi Arabia

M. Abdullah
Department of Electrical, Electronic
and Systems Engineering (EE)
Space Science Centre (ANGKASA)
College of Engineering and Built
Environment
The National University of Malaysia
Bangi, Malaysia

ISSN 2191-530X ISSN 2191-5318 (electronic)
SpringerBriefs in Applied Sciences and Technology
ISBN 978-981-16-5044-4 ISBN 978-981-16-5045-1 (eBook)
https://doi.org/10.1007/978-981-16-5045-1

This Springer imprint is published by the registered company Springer Nature Singapore Pte Ltd.
The registered company address is: 152 Beach Road, #21-01/04 Gateway East, Singapore 189721,
Singapore

Preface

Ionospheric delay is an important factor in understanding ionospheric morphology. Modelling practices ensure comparatively precise positioning, navigation, satellite communication and electromagnetic wave propagation. The investigation and forecasting of transionospheric propagation errors offer significant information as a reference in satellite and space navigation, space geodesy and radio astronomy applications. The broadcast Klobuchar model, which is the most common model used in ionospheric correction, can correct only 50% of the total effect. Thus, identifying the best approach to the ionospheric delay is necessary to effectively eliminate such effects over equatorial regions. This study focuses on an assessment and forecast of transionospheric propagation errors to attain precise measurements for improved results using a developed approach called the Holt-Winter method. This method was modified to investigate the forecasting of the ionospheric delay using global positioning system–total electron content (GPS–TEC) measurements to enhance the knowledge on the effects of ionospheric delay errors on GPS navigation in equatorial regions. The Holt-Winter method describes the periodicity of the ionospheric delay. Developed equations perform well in different regions of Malaysia during quiet periods and periods of geomagnetic disturbance. The results obtained from this study demonstrated the ionospheric delay variability, as well as the effectiveness and capability of the Holt-Winter method, which can accurately forecast the ionospheric delay and thus can be used as a regional model in equatorial regions. These results are highly beneficial in providing GPS data correction to GPS users and assisting the future development of satellite-based augmentation systems in equatorial regions.

Riyadh, Saudi Arabia   N. A. Elmunim
Bangi, Malaysia   M. Abdullah

Acknowledgements

This book was made possible with the help of God and the inspiration, guidance, comments and support of several people. I thank all of those who have contributed to my work in various ways.

Firstly, I would like to give heartfelt thanks to my beloved parents for providing consistent encouragement for my intellectual endeavours and for supporting my educational journey, while instilling in me the confidence that anything is possible with hard work. I am highly indebted to them.

I acknowledge the National University of Malaysia (UKM), Space Science Centre (ANGKASA) and National Observatory Langkawi for providing the GISTM data from the UKM and Langkawi station. Furthermore, I am grateful to the Department of Survey and Mapping Malaysia (DSMM) for providing the GPS data from the TGPG, UUMK and MUKH stations. I also thank the UKM and Malaysian Government through the Ministry of Education for the research grants (FRGS/2/2013/TK03/UKM/02/1 and GUP-2013-047) that have enabled me to pursue this project.

Contents

Abbreviations

ARIMA	Autoregressive Integrated Moving Average
ASCII	American Standard Code for Information Interchange
C/A	Coarse-Acquisition
CDMA	Code Division Multiple Access
CME	Corona Mass Ejections
COSPAR	Committee On SPAce Research
CS	Control Segment
DoD	Department of Defence
DSMM	Department of Survey and Mapping Malaysia
ECEF	Earth Centre Earth Fixed
EGNOS	European Geostationary Navigation Overlay System
ESA	European Space Agency's
GA	Ground Antennas
GBAS	Ground-Based Augmentation Systems
GISTM	GPS Ionospheric Scintillation and TEC Monitor
GLONASS	GLObal NAvigation Satellite System
GNSS	Global Navigation Satellite System
GPS	Global Positioning System
HF	High Frequency
ICO	Intermediate Circular Orbit
ICTP	International Centre for Theoretical Physics
IGAM	Institute for Geophysics, Astrophysics and Meteorology
IMF	Interplanetary Magnetic Field
IPP	Ionospheric Pierce Point
IRI	International Reference Ionosphere
ITU-R	International Telecommunications Union Recommendation
LAAS	Local Area Augmentation System
LOS	Line Of Sight
LT	Local Time
MAD	Mean Absolute Division
MAPE	Mean Absolute Percentage Error
MCS	Master Control Station

MS	Monitor Stations
MSAS	Multifunctional Satellite Augmentation System
MSD	Mean Square Division
N	Number of total observations
NAVSTAR	NAVigation Satellite Timing and Ranging
NN	Neural Networks
OCXO	Oven-Controlled Crystal Oscillator
P	P code
PIM	Parametrized Ionospheric Model
PRN	Pseudo-Random Noise
SBAS	Satellite-Based Augmentation System
SLM	Single-layer Model
SLOG	Scriptable LOgger users Guide
SSC	Storm Sudden Commencement
STEC	Slant Total Electron Content
TEC	Total Electron Content
UKM	Universiti Kebangsaan Malaysia
URSI	International Union of Radio Science
UTC	Universal Time Coordinated
VHF	Very High Frequency
VLF	Very Low Frequency
VTEC	Vertical Total Electron Content
$VTEC_x$	The measured VTEC
$VTEC_y$	The modelled VTEC
WAAS	Wide Area Augmentation System
WGS	World Geodetic System

Chapter 1
Introduction

1.1 Background

The ionosphere layer, the ionised region of the Earth's upper atmosphere, is located at altitudes ranging from 60 to 1500 km from the surface of the Earth and is regarded as a highly variable and complex physical system with sufficient numbers of free electrons that can have a considerable influence on radio wave propagation [4]. The ionosphere has a notable influence on the navigation and communication of satellites because it behaves as a dispersive medium. This influence intensifies as the density of free electrons increases and vice versa. It can alter the strength and phase of the frequency of electromagnetic radio waves. The ionosphere is used as a medium for signal transmission to satellites. Carriers experience a phase advance and codes experience group delay because electrons accumulate along the signal's path between satellites and receivers. The pseudoranges of carrier phases are drastically shorter and the pseudoranges of codes are considerably longer than the geometric range between receivers and satellites. These differences cause substantial errors in positioning and navigation systems [1].

Total electron content (TEC) is an important ionospheric parameter that measures the total number of electrons in a 1 m^2 cross section of columns along signal paths from satellites to receivers. The rapid rate of change in the TEC can create phase and amplitude delays in GPS signals. These delays can cause lock loss in GPS receivers in disturbed conditions [2]. Ionospheric influences on GPS signal propagation can have a strong effect by reducing the accuracy and reliability of GPS applications [7]. Therefore, the identification of precise models and accurate ionospheric delay forecasts is critical. Ionospheric delay changes in accordance with several factors, including the location of the user, the angle of elevation, the time of day, the seasons of the year and the solar cycle. The variations in the TEC and ionospheric delay in the equatorial region are higher than in other regions, such as mid and high latitude regions. During high solar activity, the delay can generate vertical errors ranging from 5 m and 15 m, these errors can reach 100 m under certain conditions [5].

© The Author(s), under exclusive license to Springer Nature Singapore Pte Ltd. 2021   1
N. A. Elmunim and M. Abdullah, *Ionospheric Delay Investigation and Forecasting*,
SpringerBriefs in Applied Sciences and Technology,
https://doi.org/10.1007/978-981-16-5045-1_1

Under geomagnetically disturbed conditions, charged particle concentration varies, which results in a fluctuation in TEC value. Moreover, the Earth's magnetic field is affected by solar disturbances that cause geomagnetic storms. Geomagnetic storms have initial, main and recovery phases. Geomagnetic storm conditions can induce rapid fluctuations in the TEC of GPS signals [3] that considerably affect GPS navigation performance. Therefore, the forecasting and assessment of transionospheric propagation errors, such as ionospheric delay, are important for accurate measurement because they provide relevant data for space and radio astronomy applications, space geodesy and satellite navigation. Other techniques have been developed for ionospheric forecasting purposes (see Sect. 4.1). However, corresponding research on equatorial regions has been scant. Very precise positioning with GPS requires relative positioning. Thus, an accurate model for the equatorial region is essential to improve the accuracy of GPS positioning. A study on ionospheric delay and forecasting is essential for understanding ionospheric delay behaviour and characteristics during geomagnetically differing periods and can help represent the model error of data assimilation, which is useful for improving ionospheric models. Moreover, such a study is important for identifying and selecting a suitable model to provide GPS data correction to GPS users, improve the accuracy of GPS positioning and assist the future development of satellite-based augmentation systems (SBAS).

1.2   Layout of the Book

The ionospheric delay error is the main source of error and a major concern in GPS applications because it corrupts positioning and time transfer results. Variations in the TEC and ionospheric delay are higher in the equatorial region than in other regions, such as mid and high latitude regions. The investigation and forecasting of transionospheric propagation errors are essential for precise measurement and further contributes valuable information to satellite and space probe navigation, space geodesy, radio astronomy and other applications. Currently available ionospheric models are suitable only for temperate regions and cannot achieve high accuracy. Moreover, most of these models don't take into account the periods of geomagnetic disturbance and are difficult to implement in practice. Although these models have been used for many years, none have been very successful in predicting ionospheric variability. High variations in electron density in the ionosphere result in difficulties in modelling. In general, ionospheric delay error correction for single-frequency GPS users is done by using the Klobuchar model or the broadcast model with accuracies ranging from 50% to, at most, 60% [6]. Given that the study of ionospheric delay and forecasting is a new subject in equatorial regions, particularly Malaysia, an accurate model has become essential for improving the accuracy of GPS positioning. Therefore, the study of ionospheric structure, errors and forecasting is crucial and can help in representing the model error of data assimilation, which is useful for improving ionospheric models. It is important for identifying and selecting a suitable model to provide GPS data correction to GPS users, improving the accuracy of GPS positioning and assisting the future development of SBAS.

This book described the investigation of ionospheric delay and forecasting by using the dual-frequency GPS ionospheric scintillation and TEC monitoring (GISTM) system and GPS receivers. The statistical Holt–Winter method was improved to acquire accurate forecasting results for the equatorial region.

This chapter gives a brief background and the book layout.

Chapter 2 provides an introductory overview of the GPS system and briefly describes GPS components, signal structures, observables, components and signal structure. Firstly, the combination of GPS observables is described. Then, the GPS time and coordinates are given. Finally, the error sources of GPS signals and modelling are detailed.

Chapter 3 describes essential background information on the formation and structure of the ionosphere, the TEC and the geomagnetic regions of the ionosphere. Vital information on major ionospheric variations, geomagnetic storm disturbances and delay error is discussed briefly. The observed variation in ionospheric delay and the estimation of the TEC and ionospheric delay are explained in the last section of this chapter.

Chapter 4 illustrates the main ionospheric modelling approaches. The performances of the ionospheric models in different equatorial regions are also compared.

Chapter 5 presents the Holt–Winter model. Its models, modelling aspects and developed equations are explained. Furthermore, the error measurement formulas are discussed, and the different Holt–Winter models are compared. An overview of the GISTM system and the set-up used is presented. Data recording, data processing and GISTM data analysis are explained. An investigation of the ionospheric delay and forecasting, including diurnal, monthly and seasonal variations, over the Universiti Kebangsaan Malaysia (UKM) and Langkawi National Observatory (Langkawi) stations are presented, and the errors of the model are discussed. Moreover, the diurnal variation during a geomagnetically disturbed period is described, and the ionospheric delay and forecast are investigated and estimated. The error during the disturbed period is also discussed. The statistical autoregressive integrated moving average (ARIMA) model is then explained. At the end of this chapter, a comparison of the Holt–Winter model with the ARIMA model forecast during the geomagnetically quiet and disturbed periods is presented, and the forecasted errors are analysed and compared. A summary of the work carried out is given at the end of the chapter.

Chapter 6 presents the variations in the ionospheric vertical TEC (VTEC) at different stations over Malaysia. Dual-frequency GPS receiver networks and data processing are also explained. In addition, the mathematical principles behind the VTEC and ionospheric delay analysis are explained. The methods used to estimate and calculate the slant TEC (STEC), the VTEC and ionospheric delay along with the GPS signal are subsequently described. The evaluation results of the Holt–Winter model are compared with those of the IRI-2012 ionospheric model by using IRI-2001, IRI01-corr and the NeQuick top-side options on the basis of diurnal, monthly and seasonal variations. The performance during geomagnetically quiet and disturbed periods and three intense geomagnetic storm events shown by the IRI-2016 model is compared with that of the IRI-2012 and the Holt–Winter model. At the end of the chapter, the work performed is summarised.

References

1. M.S. Bagiya, H.P. Joshi, K.N. Iyer, M. Aggarwal, S. Ravindran, B.M. Pathan, TEC variations during low solar activity period (2005–2007) near the equatorial ionospheric anomaly crest region in India. Ann. Geophys. **27**(3), 1047–1057 (2009). https://doi.org/10.5194/angeo-27-1047-2009
2. A. Coster, J. Foster, P. Erickson, INNOVATION-Monitoring the Ionosphere with GPS-Space Weather-Large gradients in the ionospheric and plasmapheric total electron content affects GPS observations and measurements. GPS Data. GPS World **14**(5), 42–49 (2003)
3. M. Förster, N. Jakowski, Geomagnetic storm effects on the topside ionosphere and plasmasphere: a compact tutorial and new results. Surv. Geophys. **21**(1), 47–87 (2000)
4. J. Klobuchar, Ionospheric effects on GPS. Glob. Positioning Syst.: Theory Appl. **1**, 485–515 (1996)
5. J.A. Klobuchar, D.N. Anderson, P.H. Doherty, Model studies of the latitudinal extent of the equatorial anomaly during equinoctial conditions. Radio Sci. **26**(4), 1025–1047 (1991). https://doi.org/10.1029/91RS00799
6. J.A. Klobucher, Ionospheric Effects on GPS, Global Positioning System: Theory and Applications, in *Fundamentals of Signal Tracking Theory 163*. ed. by B.W. Parkinson, J.J. Spilker, P. Axelrad, P. Enge (American Institute of Aeronautics and Astronautics, 1996), pp. 485–515
7. R. Warnant, I. Kutiev, P. Marinov, M. Bavier, S. Lejeune, Ionospheric and geomagnetic conditions during periods of degraded GPS position accuracy: 1. Monitoring variability in TEC which degrades the accuracy of Real-Time Kinematic GPS applications. Adv. Space Res. **39**(5), 875–880 (2007). https://doi.org/10.1016/j.asr.2006.03.044

Chapter 2
Overview of the GPS

GPS satellites orbit the Earth and transmit signals that propagate through the iono-sphere. Understanding the GPS and its components, signal structures and observ-ables, as well as the sources of errors that affect GPS signals, is crucial. Ionospheric delay is the major problem encountered in the application of GPS satellite signals. Amongst all sources of error in GPS signals, it causes the maximum errors.

2.1 Overview of the GPS

The typical generic term 'global navigation satellite system (GNSS)' is used to describe satellite navigation systems that provide autonomous geospatial positioning with worldwide coverage. GNSS enable receivers to determine their position within a few metres by using time signals that are transmitted along a line of sight (LOS) through radio signals from the satellites. Four global GNSSs are available: the GPS (United States), the GLONASS (Russia), GALILEO (E.U.), and BeiDou (China). Currently available regional systems include the Quasi Zenith Satellite System (Japan) and the Indian Regional Navigation Satellite System/NAVigation with Indian Constellation (India). The GNSS can be classified into GNSS-1 and GNSS-2. GNSS-1 is a first-generation system. It combines the GLONASS and GPS with a SBAS or a ground-based augmentation system (GBAS). The satellite-based component in the United States is the wide area augmentation system (WAAS); that in Japan is the multifunctional satellite augmentation system (MSAS), and that in Europe is the european geostationary navigation overlay system (EGNOS). Ground-based augmentation provides a system such as the local area augmentation system. GNSS-2 is a second-generation system that represents a complete civilian satellite naviga-tion system. It is epitomised by the European GALILEO positioning system. These systems provides the integrated and accurate monitoring required for civil navigation.

© The Author(s), under exclusive license to Springer Nature Singapore Pte Ltd. 2021
N. A. Elmunim and M. Abdullah, *Ionospheric Delay Investigation and Forecasting*,
SpringerBriefs in Applied Sciences and Technology,
https://doi.org/10.1007/978-981-16-5045-1_2

The GPS is an intermediate circular orbit (ICO) satellite navigation system that is used to obtain precise locations and provide an accurate time reference virtually worldwide. The GPS was developed by the department of defense (DoD) of the United States government in 1973 mainly to support soldiers and military vehicles, ships and planes for the accurate determination of their locations worldwide. The first satellite was launched in 1978 [1]. The application of the GPS has expanded to the scientific and commercial worlds. Commercially, the GPS is used for positioning and navigation in boats, airplanes and cars. It is also utilised in several outdoor recreational activities, such as hiking, jungle trekking and fishing. Its scientific applications have important roles in earth and atmospheric sciences, as well as remote sensing. The GPS is a completely functional GNSS that comprises at least 28 satellites. It is in the form of a constellation of satellites orbiting the Earth. This constellation has six Earth-centred orbital planes with four satellites in each plane. The orbits are approximately circular with similar spacing around the equator at an angle of 60° with an inclination of approximately 55° that is proportional to the equator. These satellites orbit the Earth at an altitude of approximately 20 200 km to provide users with correct time and velocity data and accurate positioning everywhere around the world and under all weather conditions 24 h per day. The satellites orbit the Earth within approximately 11 h and 58 min or half of a sidereal day. Figure 2.1 shows the constellation of GPS satellites.

Each GPS satellite uninterruptedly broadcasts the signals that are received by GPS receivers. GPSs must ensure that a minimum of four satellites are locally present above the horizon because a GPS receiver estimates its position by determining the distance between itself and four or more GPS satellites. The determination of the time delay between the reception and transmission of each GPS radio signal provides

Fig. 2.1 GPS satellite constellation (From [11])

the distance to every single satellite under the assumption that the signal travels at a velocity of 3×10^8 m/s. The signal transmits information regarding the location of a satellite. By determining its distance and position, a receiver can calculate its position on the basis of resection by distances. Although three satellites are sufficient to determine the earth centre earth fixed (ECEF) coordinates of a receiver, additional range measurement from another satellite is needed to correct the error in a receiver's clock to obtain the actual position of a receiver. Conventionally, a receiver can track range measurements from only four satellites. However, current GPS receivers can track range measurements from more than four satellites at a time (i.e. 10 to 11 satellites), thus improving positioning.

2.2 GPS Components

The GPS is a space-based system that provides positioning and timing data to users across the globe. In the last few years, the use of the GPS in monitoring iono-spheric characteristics and variations has received considerable interest. The main aim of the system is to ensure navigation, relative positioning and time transfer infor-mation worldwide. Signals from GPS satellites traverse the dispersed ionosphere, which carries its own signature, and provide outstanding opportunities for large-scale ionospheric studies across the globe. As shown in Fig. 2.2, the GPS configuration comprises three distinct segments: the user, control and space segments. All of these three segments are essential for enabling the accurate evaluation of a user's location without interruption.

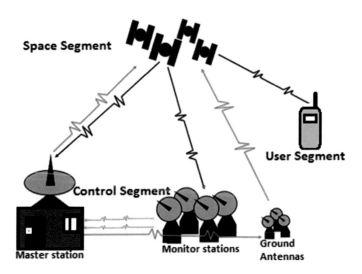

Fig. 2.2 Three GPS segments

2.2.1 Space Segment

The space segment comprises satellites and transmitted signals. Up to 32 satellites currently exist, with the exact number of satellites varying as older satellites are superannuated and replaced. The space segment is located at an altitude of approximately 20 200 km from Earth station receivers [7]. It stores data obtained from control segment stations. Additionally, it maintains accurate time by means of several onboard atomic frequency clocks. As its main task, the space segment transmits signals and information to users on one or both of the L-band frequencies. It also provides a stable platform and orbit to L-band transmitters.

2.2.2 Control Segment

The control segment (CS) is referred to as the heart of the GPS communications system. It maintains the operation of GPS on the Earth and on satellites. As shown in Fig. 2.3, the CS can be further divided into three major components: the master control station (MCS), ground antenna (GA) and monitor stations (MSs). The MCS is situated at the Schriever (previously named Falcon) Air Force Base near Colorado Springs, Colorado. It is designed to monitor and manage the satellite constellation, satellite health and maintenance activities. MSs are spread around the globe in terms of longitude. They are located at Kwajalein and Hawaii in the Pacific Ocean, Colorado Springs in Colorado, Ascension Island in the Atlantic Ocean and Diego Garcia in the Indian Ocean. Every MS checks the correct altitude, position, velocity and ephemeris of orbiting satellites. MSs track all GPS satellites to collect range data from each satellite and transmit them to the MCS [10]. The GA receives, monitors and transmits

Fig. 2.3 GPS control segment (from [2])

correction information to individual satellites. Meanwhile, the MCS uses the GA to uplink navigation data to each satellite once a day.

2.2.3 User Segment

The user segment consists of GPS receivers that can be used or installed on submarines, tanks, aircraft, ships, trucks and cars. GPS receivers detect, decode and process signals from GPS satellites. During the time of the observation, each receiver must lock onto the signals from a minimum of four GPS satellites to provide a full three-dimensional position [3]. Given that this process is passive communication wherein the receivers only receive a signal from a GPS satellite, the GPS can serve a countless number of users at a time.

2.3 GPS Signal Structure

A signal from a GPS satellite is generated by precise on-board atomic clocks (caesium or rubidium) oscillating at a fundamental frequency of 10.23 MHz with the long-term stability of 10^{-13} to 10^{14} over a day. The satellites broadcast on two basic carrier frequencies in the L band (L1 = 1575.42 MHz and L2 = 1227.60 MHz), which have the corresponding wavelengths of 19 and 24 cm and are coherently obtained by multiplying fundamental frequency with the two different integers 154 and 120, respectively, upon which binary-coded timing information signals are transmitted [1]. The GPS signal uses a code division multiple access (CDMA) technique to allow all the satellites to use the same frequency without interference. Low-rate message data based on CDMA signals are encoded with a high rate pseudo-random noise (PRN) sequence that is unique to each satellite. The receiver must be aware of the PRN code of each satellite to decode signals. The L1 carrier has two modulated codes: the lower-precise coarse-acquisition code (C/A) modulated at 1.023 MHz (chip length = 293.1 m) and the precise code (P) modulated at 10.23 MHz (chip length = 29.31 m). As shown in Fig. 2.4, the L2 carrier has only one modulated code, namely, the P-code modulated at 10.23 MHz. The C/A and P-codes are referred to as the PRN code, which when modulated on the carrier frequency gives rise to a spread spectrum signal that is resistant to jamming. In contrast to the C/A code, the P-code can be encrypted into the Y-code when the antispoofing operation mode is activated. The C/A code is repeated each millisecond, whereas the P-code repeats itself each week starting at the beginning of the GPS week (Saturday/Sunday midnight). In addition to the above two types of code, a third navigation data message with low bit rate (50 Hz) is modulated on the L1 and L2 carriers. This data message contains broadcast ephemeris (satellite orbital parameter), satellite clock adjustment and almanac data (ephemeris for all satellites), as well as ionospheric information.

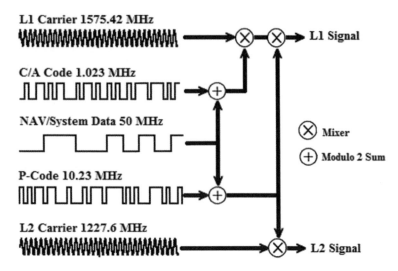

Fig. 2.4 GPS satellite signals (From [2])

The modernised GPS signal includes three new civilian signals, namely, L2C, L5 and L1C, and a new military signal M on the L1 and L2 frequencies. The simultaneous application of the L2C (second civilian) and L1 C/A (first civilian) in a dual-frequency receiver allows for ionospheric correction, which in turn increases the accuracy of the signal compared with that of the military signal. In this work, the L1 and L2 frequencies were used.

2.4   GPS Observables

GPS observables have two fundamental signals: the pseudorange (code) and the carrier phase. The carrier phase is favoured, although pseudoranges are often used in navigation and high-precision positioning. Pseudoranges and the carrier phase can be used in ionospheric modelling. Phase observation has the advantage of low noise. However, for each satellite, at least one extra unknown parameter, which is integer ambiguity, has to be introduced. The pseudorange or biased range measures the distance between a satellite and a receiver and refers to the epochs of the transmission and reception of the codes scaled by the nominal speed of light in a vacuum. Unavoidable timing errors or offsets concerning the GPS time in the satellite and receiver clocks and the delays due to the atmosphere and ionosphere differentiate the measured pseudoranges from the LOS, which corresponds to the epochs of transmission and reception. The mathematical expression for the pseudorange is given as

$$P_r^s = c(t_r - t^s) = c\tau_r^s = \rho_r^s + c(\delta t_r - \delta t^s) + I + T + mp + \varepsilon \qquad (2.1)$$

where

P_r^s	pseudorange measured at the receiver.
c	speed of light in a vacuum.
t^s	transmission time of signal measured by the time frame of satellite s.
t_r	reception time of signal measured by the clock of receivers r.
τ_r^s	signal traveling time.
ρ_r^s	LOS range from satellite antenna and receiver antenna.
$\delta t_r, \delta t^s$	satellite and receiver clock error due to the difference in system time.
I	ionospheric-induced error.
T	tropospheric-induced error.
mp	multipath error.
ε	noise or random error.

The carrier phase, or ambiguous range, is equal to the difference between the phase of the carrier signal generated by the receiver at the signal reception time and the phase of the satellite signal generated at signal transmission time. The initial observation only entails the fractional part of the phase difference. The receiver continues recording this fractional part along with the whole number of cycles since the initial epoch when tracking continues without a cycle slip. However, this measurement does not provide the initial integer number, thus introducing ambiguity. The carrier phase is measured normally in the unit of cycles and can be scaled to units of length through multiplication by wavelength at the operational frequencies ($\lambda = c/f$). Thus, the phase equation in units of length is defined as

$$\phi_r^s = \rho_r^s + c\left(\delta t_r - \delta t^s\right) - I + T + \lambda N_r^s + mp + \varepsilon \qquad (2.2)$$

where

ϕ_r^s	phase measurement in units of length.
N_r^s	integer ambiguity between a satellite and a receiver.
$\delta t_r, \delta t^s$	satellite and receiver clock error due to the difference in system time.
I	ionospheric-induced error.
T	tropospheric-induced error.
mp	multipath error.
ε	noise or random error.

Orbital error, antenna phase centre offset and receiver noise are noises or error sources that may be included in both observables and must be considered if an accurate measurement is required. Notably, the opposite sign of the ionospheric error between Eqs. (2.1) and (2.2) is due to transionospheric propagation effects ($I_p = - I_\phi$). The magnitude of the ionospheric-induced group delay is identical to the phase advance when path bending and the geomagnetic field effect are ignored. Phase observations are two/three orders of magnitude more precise than the pseudorange code but contain ambiguities. The LOS range from the satellite coordinates (X^s, Y^s, Z^s) to the receiver's coordinates (x_r, y_r, z_r) is given as.

Table 2.1 Main characteristics of the carrier phases and code

	Code	Carrier
Wavelength	P-code 29.3 m C/A-code 293 m	L1 19.0 cm L2 24.4 cm
Observation noise	P-code 0.3–1 m C/A-code 3–10 m	1–3 mm
New development	P-code cm–dm C/A-code dm–m	<0.2 mm
Propagation effect ambiguity	Ionospheric delay effects + I nonambiguous	Ionospheric advance—I ambiguous

$$\rho_r^s = \sqrt{(X^s - x_r)^2 + (Y^s - y_r)^2 + (Z^s - z_r)^2}. \tag{2.3}$$

Both observations represent the fundamental GPS observables. Their main characteristics are listed in Table 2.1.

2.5 GPS Time and Coordinates

GPS satellites use precise and stable caesium atomic clocks. However, GPS time is not synchronised to the universal time coordinated (UTC). It has no leap seconds or other corrections necessary to synchronise with the UTC. The MCS is used to monitor and maintain the correction of this clock, which is periodically added to the UTC to synchronise with GPS time. The MCS creates clock correction parameters for transmission to satellites through the navigation data that are received by control stations [6]. Then, these navigation data, which determine the precise magnitude of clock offsets from the UTC, are retransmitted to GPS users. The recent version of the UTC is described by the International Telecommunications Union Recommendation (ITU-R TF.460–6) and standard-frequency and time-signal emissions in accordance with the International Atomic Time with the addition of leap seconds at inconsistent intervals to compensate for the slow rotation of the Earth. Leap seconds maintain the UTC by approximately 0.9 s. A total of 18 leap second additions have been made with the most current addition added on 30 December 2016.

The World Geodetic System 1984 (WGS 84) is used to determine GPS satellite orbits. The WGS 84 gives a reference frame for the Earth to be utilised in geodesy and navigation on the basis of geocentric Cartesian x, y and z coordinates. These Cartesian coordinates are known as ECEF and are expressed in metre units. Users determine their location in this coordinate system before transforming them into geodesic coordinates (latitude, longitude and altitude).

2.6   Error Sources in GPS Signals and Modelling

Several error sources can reduce the accuracy of location determination by a GPS receiver. These sources of error can be categorised into three major groups: (1) satellite-based error, which includes ephemeris and satellite clock biases; (2) receiver-based error, which comprises receiver clock errors, interchannel biases and receiver noise and (3) propagation error, which consists of ionosphere and troposphere delay, multipath signals and other interferences. The possible error sources of GPS measurements are shown in Fig. 2.5.

The travel time of GPS satellite signals can be altered by atmospheric influence. The refraction of a GPS signal when it is transmitted through the troposphere and the ionosphere leads to differences in signal speed from the GPS signal speed in space. Sunspot activity also results in interference with GPS signals. Measurement noise is another notable source of error. Errors also occur via signal deformation resulting from electrical interference or are peculiar to the GPS receiver itself. Inconsistency in ephemeris data (the satellite orbit data) can also lead to errors in the determined positions because the satellites are not present where the GPS receiver estimates positions in accordance with received information. A small difference in the atomic clocks aboard satellites can translate into large positioning errors; a clock error of 1 ns is equivalent to a 3 m user error on the ground. Multipath effects occur when transmitted signals from the satellites rebound from a reflective surface before reaching the

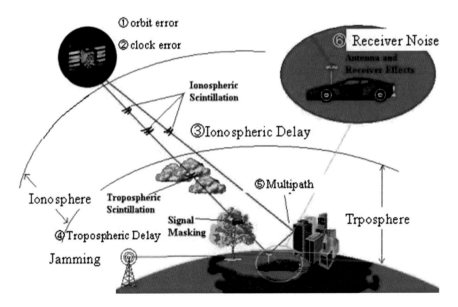

Fig. 2.5   Error sources in GPS signals (From [5])

Table 2.2 Summary of GPS error sources

Error source	Potential error	Typical error
Ionosphere	5.0 m	0.4 m
Troposphere	0.5 m	0.2 m
Ephemeris data	2.5 m	0 m
Satellite clock drift	1.5 m	0 m
Multipath	0.6 m	0.6 m
Measurement noise	0.3 m	0.3 m
Total	~ 15 m	~ 10 m

receiving antenna. Given this situation, the receiver receives the signal in a straight-line path and a delayed path (multiple paths). These error sources contribute to the error sources listed in Table 2.2 and are explained briefly as follows:

2.6.1   Random Noise Measurement

Code delay and carrier phase measurements are affected by receiver noise, which is termed random measurement noise. This noise and the signal are related. Noise emanates from the antenna, cables, amplifiers and from the receiver itself. Interference from other GPS signals is also included in random measurement noise. Although current GPS receiver technology has been upgraded to eliminate this error, it also contributes a small amount of error in the overall positioning.

2.6.2   Multipath

The multipath effect is a highly localised phenomenon that occurs when a received GPS signal contains a direct LOS signal and one or more indirect signals that have been reflected by objects in the receiver's vicinity. Multipath effects depend on a number of factors, including satellite geometry, the presence of reflective surfaces around the receiving antenna and the position of the antenna itself. From a purely geometrical standpoint, signals that are received from satellites at a low elevation angle are more prone to multipath effects than signals received from a high elevation angle. Multipath effects can be considered as systematic errors over short time spans (several minutes). Their effect on pseudorange measurement is substantially greater than on the carrier phase, wherein P-code multipath observational errors are two orders of magnitude larger than carrier phase observational errors, which can reach the decimetre to metre level. Multipath effects on carrier phase measurements introduce a phase shift that degrades the range measurement by several centimetres because of periodic bias. The maximum error on the L1 carrier signal due to a phase shift of 90 ° corresponds to approximately 5 cm. Given that satellites change geometry in the sky, the multipath effect introduces a cyclical error into the carrier phase observations with a classical period ranging from 15 to 30 min depending on site characteristics.

2.6.3 Clock Error

Clock errors can be attributed to the drifts amongst the real GPS time, the satellite clock time and the true GPS time. The transmitted navigation message from satellites includes corrections for clock drifts. However, some corrected clock irregularities result in positional errors. Satellite clock errors are normally less than 1 ms, which is equal to a 300 km pseudorange error. They can be eliminated by differentiating the observations between two different receivers on the basis of their respective satellites. Meanwhile, receivers use cheap crystal clocks that have lower accuracy than the atomic clocks on satellites. Hence, receiver clock errors are considerably larger than satellite clock errors. The magnitude of a receiver clock error ranges from 200 ns to several ms and depends on the internal firmware of the receiver. This error can be eliminated by differentiating the observations of the same receiver between two different satellites [9].

2.6.4 Ephemeris Error

Satellite ephemeris errors are errors due to the incorrect prediction of satellite positions, which may then be transmitted by satellites to users as a part of the navigation message. The position of the satellite is dynamic and is a function of the gravitational field and solar pressure. The computation of position by a ground MCS is subject to error as a result of clock drifts and processing delays in the monitoring stations. Errors in the estimation of satellite position culminate in pseudorange errors and need to be corrected at the user level.

2.6.5 Tropospheric Delay

The troposphere is the lowest region in the atmosphere. It expands from the surface of the Earth to a height of approximately 18 km at the equator and 6 km at the South and North Poles of the Earth. All weather phenomena occur in this layer. When GPS signals propagate via the troposphere, they suffer from the effects of tropospheric attenuation. Given that the medium is nondispersive at GPS frequencies, tropospheric error cannot be eliminated by using a dual-frequency receiver as is the case for ionospheric error. Tropospheric errors are categorised into two components, i.e. dry and wet, on the basis of the magnitude of delay errors. Although the dry component contributes 80%–90% of delays, it can be modelled with an accuracy of less than 1% of the overall tropospheric error. The wet part, which contributes only 10–20% of the delay, is more unpredictable and difficult to model than the dry part [4].

2.6.6   Ionospheric Delay

GPS signals are nonlinearly dispersed and refracted mostly by free electrons in the ionosphere. The dispersive nature of the ionosphere allows the delays to become frequency-dependent. Hence, compared with lower frequencies, higher frequencies experience lesser delays, i.e. GPS L1 (1575.42 MHz) has a lesser ionospheric delay than L2 (1227.60 MHz). The presence of the ionosphere increases the velocity of the carrier phase to above the speed of light but delays PRN codes and the navigation message by a similar amount. Additional effects include range-rate errors and scintillations in GPS radio signals. The total ionospheric delay is proportional to the number of free electrons through the signal ray path, i.e. the TEC, which varies with the time of day, year, solar cycle and geographic location on or over the Earth. In the case of a single-frequency GPS receiver, only approximately 50–60% of the ionospheric error can be minimised by using the Klobuchar model, whose coefficients (eight parameters) are transmitted as part of the navigation message [8]. In addition, single-frequency GPS users may apply empirical ionospheric models or real-time correction from the regional network through communication links. Given that the ionosphere is the major source of errors in GPS measurements, its properties and variations and the identification of accurate ionospheric delay approaches are the major focus of this work.

References

1. N. Ashby, J.J. Spilker, Introduction to Relativistic Effects on the Global Positioning System. Glob. Positioning Syst.: Theory Appl. **1**, 4–17 (1996)
2. P.H. Dana, Global positioning system (GPS) time dissemination for real-time applications. Real-Time Syst. **12**, 9–40 (1997). https://doi.org/10.1023/A:1007906014916
3. P. Enge, P. Misra. (2011). Global positioning system: signals, measurements, and performance - revised second edition (2011). Int. J. Wireless Inf. Netw. (Issue 2). http://www.navtechgps.com/assets/1/7/2500-2.pdf
4. G. Fotopoulos, M.E. Cannon, G. Fotopoulos, M.E. Cannon, An overview of multi-reference station methods for cm-level positioning. GPS Solutions **4**(3), 1–10 (2001). https://doi.org/10.1007/PL00012849
5. GPS Limitations. (n.d.). *The BlackBoxCamera Company Limited - GPS display unit*. Retrieved January 3, 2017, from http://www.blackboxcamera.com/pic-osd/gps_limits.htm
6. B. Hofman-Wellenhof, H. Lichtenegger, J. Collins. *Global Positioning System: Theory and Practice*. (Springer Wien, New York, 1993)
7. E.D. Kaplan, *Understanding the GPS: Principles and Applications*. (Artech House, 1996)

8. J.A. Klobuchar, D.N. Anderson, P.H. Doherty, Model studies of the latitudinal extent of the equatorial anomaly during equinoctial conditions. Radio Sci. **26**(4), 1025–1047 (1991). https://doi.org/10.1029/91RS00799

9. G. Lachapelle, GPS Theory and Applications. In *ENGO625*, (1998)

10. R.J. Norman, P.S. Cannon, An evaluation of a new two-dimensional analytic ionospheric ray-tracing technique: segmented method for analytic ray tracing (SMART). Radio Sci. **34**(2), 489 (1999). https://doi.org/10.1029/98RS01788

11. S. Pal, A. S. Ganeshan (n.d.), *GNSS: The Technological Leap—Geospatial World*. Retrieved January 24, 2017, from https://www.geospatialworld.net/article/gnss-the-technological-leap/

12. The BlackBoxCameraTM Company. *The BlackBoxCamera Company Limited—Precision GPS time overlay unit* (2016). http://www.blackboxcamera.com/pic-osd/sprite.htm

Chapter 3
Ionosphere

The ionosphere is a notable source of error that disrupts the accuracy of GPS signals propagating to the ground by changing the velocity of the propagated signals. This effect causes a delay in GPS positioning. Therefore, studying the ionosphere is crucial for eliminating ionospheric delay. This chapter provides an explanation of ionospheric formations and structures, as well as ionosphere variation, which carries great importance for modelling and forecasting delay in equatorial regions. Ionospheric electron density shows substantial changes when the effects of ionospheric geomagnetic disturbances are considered. The understanding and estimation of the TEC and ionospheric delay are important for investigating ionospheric delay and identifying an accurate forecasting model for a specific region.

3.1 Formation of the Ionosphere

The ionosphere is a part of the upper atmosphere that extends from 60 to 1500 km above the Earth's surface. The main constituents of the ionosphere are ionised gases that are produced from solar radiation. Ultraviolet radiation from the sun excites atmospheric molecules to produce free electrons and ions [3]. The main influence of the ionosphere on GPS signals is due to these free electrons. Free electrons change the propagation velocity of signals from GPS satellites, thus generating ranging errors. Electron density depends on solar radiation intensity, which is a function of several factors, including local time, season and solar cycle. During the daytime, high solar radiation breaks down molecules and produces numerous free electrons. At night, without further ionisation, free electrons recombine with ions, thus reducing electron density. The seasonal variation of electron density is due to the Earth's revolution around the sun, and the angle of solar radiation changes with the season. Electron density also varies with the 11-year solar cycle that is intimately related to the number of sunspots. The solar cycle is a measure of the periodic variation of the sun's activity.

© The Author(s), under exclusive license to Springer Nature Singapore Pte Ltd. 2021   19
N. A. Elmunim and M. Abdullah, *Ionospheric Delay Investigation and Forecasting*,
SpringerBriefs in Applied Sciences and Technology,
https://doi.org/10.1007/978-981-16-5045-1_3

During the maximum of the solar cycle, the increased level of solar radiation produces additional free electrons in the ionosphere. The ionosphere is separated into three main layers as explained in detail in Sect. 3.2.

3.2 Structures of the Ionosphere

The ionosphere can be divided into the D, E and F layers on the basis of electron density.

Figure 3.1 presents the D, E and F (F1 and F2) layers. This figure shows a view of the Earth from the top of the north geographic pole. During the daytime, the radiation of the sun on the local atmosphere is high, and all of the layers are present. Ions and electrons recombine during the night-time or in the absence of solar radiation. Negatively charged electrons and positively charged ions combine to produce neutral atoms. In the late afternoon and early hours of the evening, the recombination rate is higher than the rate of ionisation. Consequently, the density of the ionised layers normally starts decreasing in the D, E and F1 layers during this period. The F2 layer does not behave similarly. The F2 layer electron density attains its lowest value slightly before the night-time. Only the F2 layer is visible because the D, E and F1 layers nearly disappear completely as recombination proceeds. The F2 layer is simply called the F layer. Then, as the sun rises, photoionisation occurs, leading to a continuous increase in the free electron density in the F layer. This phenomenon makes the ionosphere useful for the high-frequency (HF) or ionospheric reflected propagation of radiowaves during the night-time. Further discussion on the layers is given below.

Fig. 3.1 A simplified view of ionosphere layers during the day and night (From [8])

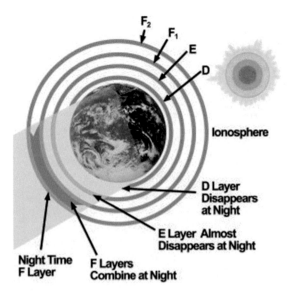

3.2.1 D Layer

The D layer is located at an altitude of approximately 60–90 km above the Earth's surface. It is the lowest layer of the ionosphere. In radio wave propagation, the D layer can absorb and attenuate the energy of radio waves at MF, HF and VHF while reflecting LF and VLF signals. This layer is present only during the hours when the sun is locally incident on the ionosphere at the height of the D region. The rapid recombination of ions after sunset leads to the disappearance of the D layer and is the main reason why the D layer has little effect on HF radio wave propagation during the night-time.

3.2.2 E Layer

The E layer of the ionosphere is located approximately 90–130 km above the surface of the Earth. Electron density is higher at lower latitudes and is also greater during the daytime than during the night-time, during which it almost disappears. Sporadic E (sometimes designated as Es) is another layer that can exist in this region at an altitude of approximately 90–120 km or more. In contrast to the normal critical frequency of the E layer, the critical frequency of the E layer is highly variable in time and space.

3.2.3 F Layer

The F layer is the most important ionosphere layer for HF and transionospheric radio wave propagation. It exists from an altitude of approximately 130 km up to 500 km. During the daytime, the F layer divides into the F1 and F2 layers. The F1 layer extends from altitudes of approximately 130 km to 210 km and disappears during the night due to dissociative recombination. The F2 layer extends from an altitude of approximately 210–400 km or sometimes even up to 500 km and is always present throughout the daytime and night-time. During the daytime, the equatorial and low-latitude ionospheres exhibit electron density at the F region at the geomagnetic equator and two crests within the ± 20° magnetic latitude. In particular, the density at the F region represents the largest constituent of the maximum TEC in the ionosphere that can substantially affect the propagation of radio waves.

3.3 Total Electron Content

The TEC is measured by the quantity of free electrons in a cylinder of unit cross-section via the signal path that extends from a satellite to a receiver on the ground and is inversely proportional to the square of the frequency of transionospheric radio wave. It considerably affects the electromagnetic waves that propagate through the ionosphere. GPS-based ground and navigation positioning that utilises transionospheric communication has promoted the study on the TEC. Given that ionospheric TEC highly depends on the state of the ionosphere, it has a domineering effect on GPS based communications. Ionospheric TEC substantially varies with geographical location (high-latitude, midlatitude, low-latitude and equatorial regions). This phenomenon is highly obvious in low-latitude and equatorial regions, such as Malaysia, because of the unique geometry of the magnetic field. In such regions, plasma density in the ionosphere causes substantial variations in the TEC as a function of season, latitude, longitude, time of the day and solar and geomagnetic activity.

3.4 Geomagnetic Regions of the Ionosphere

As illustrated in Fig. 3.2, the ionosphere is divided into three major regions: the high-latitude, midlatitude and low-latitude and equatorial regions. These regions are regulated by various physical processes.

The high-latitude or polar region consists of the auroral zone and is located approximately between from 60° to 70° magnetic latitude. The polar cap is located at less than 70° geomagnetic latitude. Electron density in this region is significantly lower than that in low-latitude regions because solar radiation hits the atmosphere evasively.

The mid-latitude region extends from approximately 20° to 60° on either side of the geomagnetic equator. In the mid-latitude region, ionisation is due merely to solar photon radiation and electron density is not subject to any particle radiation. Most

Fig. 3.2 Ionosphere regions (From [4]

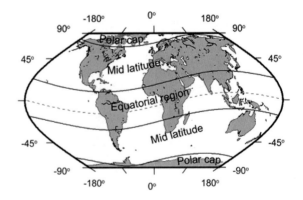

of the instruments used in observing the ionosphere are located in this region, where day-to-day variation is low.

The low-latitude and equatorial region span approximately $\pm 20°$ on either side of the geomagnetic equator. It is regarded as the region with the highest value of peak electron density because it receives high radiation from the sun in addition to the electric and magnetic fields of the earth. Accordingly, the accumulated electrons in this region influence radio wave propagation, which drastically affects satellite communication systems.

3.5 Major Ionospheric Variations

The long- and short-term variations within the ionosphere are monitored by studying the temporal and spatial distribution of electron density over an extended period of time. Most of the factors that describe ionospheric variability are formed by radiation from the sun. This formation mechanism results in variations in the ionosphere with the seasons, time of the day and position on the surface of the Earth. Ionospheric variation is also well known to follow an 11-year solar activity cycle [9].

3.5.1 Solar Cycle Variations

The sun is the major source of energy that leads to photoionisation, thereby contributing greatly to electron density within the ionosphere. Its approximately 11-year cycle includes periods of high and low solar activity. The TEC depends on the activity of the sun, and its maximum values are mainly observed during periods of high solar activity and least observed during periods of low solar activity. Sunspot number is one of the indicators that are commonly used to determine solar activity.

3.5.2 Diurnal Variations

Diurnal variations in the ionosphere simply refer to the day–night electron density variations resulting from the revolution of the Earth with respect to the sun. Electron density peaks during the daytime and decreases in the night-time given the lack of photoionisation as a result of the absence of solar radiation. In general, when photoionisation ceases, recombination greatly reduces electron density during the night-time, but some free electrons remain until dawn.

3.5.3   Seasonal Variations

Seasonal variations are simply referred to as day-to-day variations through the whole year. Electron density is lower in the winter than in the summer due to decreased photoionisation levels. Thus, the behaviour of the TEC is expected to be similar to that of electron density. Electron density is higher during equinoxes than other seasons because the sun is directly overhead the equator during periods of maximum and minimum solar activities, leading to an increase in electron density.

3.5.4   Geomagnetic Activity Effects

Ultraviolet radiation from the sun is the source of thermal convection at ionospheric heights, leading to the movement of ions and electrons across the geomagnetic field. The generated ionospheric current gives rise to a magnetic field around the ionosphere, the variations of which are later observed as geomagnetic field fluctuations on the Earth's surface. The ionospheric electron density during periods of geomagnetic storms is characterised mainly by the negative and positive effects of the storms and is dependent, amongst others, on the latitudes and the strength of the disturbances. A detailed discussion of this phenomenon is given in Sect. 3.6.

3.5.5   Latitudinal Variations

Ionospheric densities over the equator are expected to be higher than those over other geographical locations. As the latitude increases on either side of the geographic equator, the solar radiation arrives in the atmosphere at an oblique angle, resulting in the reduction in ionospheric densities. Given that solar zenith angle (ZA) determines ionisation levels, geographical locations with low ZA are exposed to high solar radiation and hence have higher electron densities than other geographical locations.

3.6   Ionospheric Geomagnetic Storm Disturbances

Corona mass ejections (CMEs) that enter the magnetic field of the Earth can lead to geomagnetic storms that usually occur together with ionospheric storms. CMEs result from solar wind eruption from the active parts of the sun. When the solar wind encroaches on the magnetosphere, its pressure and velocity, as well as embedded interplanetary magnetic field (IMF), vary widely because it originates from a different solar active region. Parameters undergo a sudden change because of the manifestation of shock waves and the southward turning of the IMF Bz component, which leads to a rapid enhancement in magnetic reconnection at the magnetopause. At a highly negative Bz, the magnetic reconnection between the geomagnetic field and IMF generates open field lines that enable momentum, energy and mass transfer from the solar wind to the Earth's magnetosphere [3]. A geomagnetic storm is described by the changes in the disturbance storm time (Dst) index. The Dst index is used to determine the global average changes of the horizontal component of the Earth's magnetic field at the magnetic equator in accordance with measurements from a few magnetometer stations. It is basically used to define the various storm phases when studying the global activities as the geomagnetic storm proceeds. The planetary K (Kp) index is used to measure and quantify the disturbances in the horizontal component of the earth's magnetic field with an integer that ranges from 0 to 9. The geomagnetic storm is indicated when the Kp is equal to or more than 5 Kp. Three different geomagnetic storm phases exist: initial, main and recovery phases. The initial phase is characterised by a 20–50 nT increase in the Dst within tens of minutes. The initial phase is also referred to as sudden storm commencement (SSC). However, some geomagnetic storms lack an initial phase, and some sudden increases in the Dst are unaccompanied by a geomagnetic storm. The main phase of a storm is characterised by the build-up of an escalated ring current by highly energetic particle injection and activation. This event is attributed to a decline in Dst values to a minimum value of less than − 50 nT. This decline marks the end of the main phase and initialises the storm recovery phase. In general, the main phase lasts for 2–8 h. The recovery phase occurs as the Dst changes from the minimum to the quiet time value. This phase can last for a minimum of 8 h or a maximum of 7 days [6, 10, 13]. Geomagnetic storms can be categorised as moderate (-50 nT > minimum Dst < -100 nT), intense or strong (-100 nT > minimum Dst < -250 nT) and extreme or superstorm (minimum of Dst > -250 nT) [5]. Several communication and navigational systems utilise radio wave signals that propagate through or reflect from the ionosphere. GPS radio signals propagate through the ionosphere and can be influenced by space weather phenomena and changes in electron density because space weather activity may affect the travel speed of radiowaves and introduce propagation delay into GPS signals. Sudden changes in electron density that result from space weather disturbances, such as geomagnetic storms, cause ionospheric delay errors that greatly affect GPS measurements. These effects lead to inaccurate positioning measurements.

3.7 Ionospheric Delay Error

Ionospheric errors in signals received from GPS satellite signals play the greatest role in the positioning errors encountered when using single-frequency GPS receivers. These phenomena become increasingly obvious during periods of severe solar activity, such as geomagnetic storms. For any radiometric space method, such as methods using GNSS, providing an account of the propagation delay that emanates from the neutral atmosphere is imperative. A radio signal is refracted when it propagates in the atmosphere due to the neutral nature of the atmosphere. Its path changes, i.e. the signal does not move in a straight line between the receiver and the signal source. Its speed also changes, i.e. the signal travels at a speed slower than the vacuum light speed. As a result of alterations in velocity, the signal takes a longer time to arrive at the receiving antenna when travelling in the atmosphere than when traveling in a vacuum. This change in time, which can also be expressed in units of length, is called the neutral ionospheric propagation delay. The effect of the ionosphere is a vital source of error in GPS measurements. The amount of ionospheric delay or the advance of the GPS signal may range from a few metres to over 20 m per day [12]. In general, the modelling of ionospheric effects is difficult because the physical interaction amongst solar activities and geomagnetic field are complicated. The ionosphere is a dispersive medium, i.e. the effect of the ionosphere depends on frequency. Consequently, the GPS is designed with various frequencies to measure or correct the effect of the ionosphere. One of the largest sources of error in a single-frequency GPS is the ionospheric effect, which refracts and slows down GPS signals when they pass through it. This propagation delay varies and depends on a number of factors, including solar activity, time of day, geographical locations (latitude) and seasons. The effect is proportional to the TEC and the ionospheric electron density line integral. Hence, a precise ionospheric model, which will provide an accurate forecast of the ionospheric delay, is crucial.

3.8 Estimation of the TEC and Ionospheric Delay

Under the assumption that a horizontal gradient does not exist in the ionosphere and that the ionosphere is a thin shell at a height referred to as the shell height, the STEC is normally mapped vertically for the TEC to be independent of the relative locations of a satellite and a receiver. The shell height is the particular height at which electrons are equally distributed above and below. It is near the peak height of the ionosphere F layer that is assumed to be located at an altitude of 450 km. The point at which the ray path crosses the thin shell is referred to as the ionospheric pierce point (IPP) (Fig. 3.3). The IPP is the point at which the entire TEC of a certain ray path is regarded as concentrated. The STEC at the IPP is subsequently mapped to the vertical by using a mapping function.

The cosecant mapping function is dependent on the height of the single-layer ionosphere. [11] showed that the mapping function is sensitive to the mean height of the ionosphere only through large zenith angles (approximately X' >70°). In addition, for large zenith angles, the TEC can be up to three times the value of the VTEC. Notably, the subionospheric point can be distant from the receiving end for approximately several thousand kilometres (for large satellite zenith angles). This situation indicates that with horizontal electron density gradients, the usage of a simple cosecant mapping function to convert the STEC into the VTEC results in considerable TEC conversion error. STEC values can start from 1 TECU to reach several hundred TECU through the signal path depending on elevation angle, time and space. In fact, variations in the TEC in time and space are mainly caused by

Fig. 3.3 Ionospheric
single-layer model

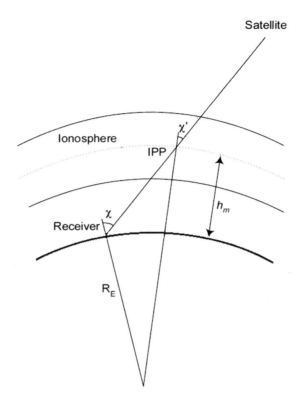

the spatial and temporal variations of ionospheric electron density. The STEC is set
to be equivalent to the raw STEC minus the receiver and satellite bias to correct
for satellite and receiver bias. Such information is available online at http://aiuws.
unibe.ch/aiub/bswuser/opb (Eq. 3.2). This approach is satisfactory for investigating
and forecasting ionospheric delay variations. When such biases are absent, VTEC
values may differ from absolute values but the observed trends remain the same [2].
However, these biases are not enhanced by ionospheric delay. Single-layer ionosphere
approximations are normally used and the STEC is expressed as an equivalent VTEC
to investigate the solar cycle and the seasonal and diurnal behaviour of the TEC by
using the thin-shell model approach to take the projection from the slant to the
vertical as proposed by Manucci et al. [7]. The converted VTEC in metres is called
the vertical ionospheric delay [1]. The VTEC is easier to compare or model than the
STEC at various elevation angles.

$$VTEC = STEC(cos\chi')$$ (3.1)

$$STEC = STEC_{raw} - B_s - B_r$$ (3.2)

$$\sin \chi' = \frac{R_E}{R_E + h_m} \sin \chi \tag{3.3}$$

$$\chi' = \arcsin = \left[\frac{R_E}{R_E + h_m} \sin \chi \right] \tag{3.4}$$

where

χ' zenith angle at the IPP.
B_r receiver bias.
B_s satellite bias.
χ zenith angle at the receiver position.
R_e mean earth radius, 6371 km.
h_m mean height of the ionosphere SLM, assumed 450 km.

A dual-frequency GPS receiver is used to measure the difference in the ionospheric delay between L1 and L2 signals. The ionospheric delay for dual-frequency GPS receivers is expressed as

$$I = 40.3 \, VTEC \left(\frac{1}{f_2^2} - \frac{1}{f_1^2} \right) \tag{3.5}$$

where

I ionospheric delay.
f_1 carrier frequency is 1575.42 MHz.
f_2 carrier frequency is 1227.60 MHz.

Applying the above ionospheric delay equation shows that an increment of 1 TECU corresponds to 0.16237 m at L1 and 0.26742 m at L2, whereas the L1, L2 combination corresponds to 0.10505 m as presented in Table 3.1.

If the GPS frequency delay values are known, then they can be added to the VTEC and the ionospheric delay can be forecasted. Otherwise, the VTEC is forecasted directly. In this book, the Holt–Winter model is used under both conditions to estimate ionospheric delay and VTEC variations and to understand the capability of the method.

Table 3.1 Ionospheric error corresponding to 1 TECU increment (>90% of the total)

GPS frequency	Delay Error (ns)	Delay error (m)
L1 (1575.41 MHz)	0.54124	0.16237
L2 (1227.60 MHz)	0.89139	0.26742
L2–L1	0.35015	0.10505

References

1. O.E. Abe, X. Otero Villamide, C. Paparini, S.M. Radicella, B. Nava, M. Rodr??guez-Bouza, Performance evaluation of GNSS-TEC estimation techniques at the grid point in middle and low latitudes during different geomagnetic conditions. J. Geodesy, (2016) 1–9. https://doi.org/10.1007/s00190-016-0972-z

2. V. Chauhan, O.P. Singh, A morphological study of GPS-TEC data at Agra and their comparison with the IRI model. Adv. Space Res. **46**(3), 280–290 (2010). https://doi.org/10.1016/j.asr.2010.03.018

3. C.J. Davis, M.N. Wild, M. Lockwood, Y.K. Tulunay, Ionospheric and geomagnetic responses to changes in IMF B z : a superposed epoch study. Ann. Geophys. **15**(2), 217–230 (1997). https://doi.org/10.1007/s00585-997-0217-9

4. M.M.A. Elizei, *Multi-dimensional modeling of the ionospheric parameters, using space geodetic techniques.* Veinna University of Technology (2013)

5. M. Förster, N. Jakowski, Geomagnetic storm effects on the topside ionosphere and plasmasphere: a compact tutorial and new results. Surv. Geophys. **21**(1), 47–87 (2000). 10.1023%2FA%3A1006775125220

6. S. Inyurt, Modeling and comparison of two geomagnetic storms. Adv. Space Res. **65**(3), 966–977 (2020). 10.1016/j.asr.2019.11.004

7. A.J. Manucci, B.D. Wilson, C.D. Edwards, A new method for monitoring the Earth's ionospheric total electron content using the GPS global network. *ION GPS-93* (1993), pp. 1323–1332

8. NASA, *Earth's Atmosphere, Schumann Resonance and the Ionosphere - HeartMath Institute.* https://www.heartmath.org/gci-commentaries/earths-atmosphere-schumann-resonance-and-the-ionosphere/(2009)

9. H. Rishbeth, M. Mendillo, Patterns of F2-layer variability. J. Atmos. Solar Terr. Phys. **63**(15), 1661–1680 (2001)

10. B.T. Tsurutani, D.L. Judge, F.L. Guarnieri, P. Gangopadhyay, A.R. Jones, J. Nuttall, G.A. Zambon, L. Didkovsky, A.J. Mannucci, B. Iijima, The October 28, 2003 extreme EUV solar flare and resultant extreme ionospheric effects: Comparison to other Halloween events and the Bastille Day event. Geophys. Res. Lett. **32**(3) (2005)

11. A.J. Van-Dierendonck, J. Klobuchar, Q. Hua, Ionospheric Scintillation Monitoring Using Commercial Single Frequency C/A Code Receivers. *Proceedings of the 6th International Technical Meeting of the Satellite Division of The Institute of Navigation (ION GPS 1993* (1993), pp. 1333–1342

12. G. Xu, Y. Xu. Applications of GPS Theory and Algorithms. In *GPS.* (Springer, Berlin Heidelberg, 2016)

13. B. Zhao, W. Wan, L. Liu, Z. Ren, Characteristics of the ionospheric total electron content of the equatorial ionization anomaly in the Asian-Australian region during 1996–2004. Ann. Geophys. **27**(10), 3861–3873 (2009). https://doi.org/10.5194/angeo-27-3861-2009

Chapter 4
Ionospheric Modelling and Forecasting

In this chapter, various ionospheric modelling and forecasting techniques are explained to understand their differences, as well as their capability to forecast and modelling the ionospheric delay. Several modelling approaches that use data collected over several years to forecast ionospheric delay are introduced. However, most of these models provide accuracies ranging from 50 to 60% or difficult to implement. The IRI model has undergone considerable progress recently and provides accurate results for midlatitude regions. Time-series models are currently used to forecast ionospheric delay. Most of these models can forecast the ionospheric delay during geomagnetic quiet periods only.

4.1 Ionospheric Modelling and Forecasting Approaches

The effect of the ionosphere on radio wave propagation has been of considerable interest since the advent of radio and satellite communications and has recently become a concern of GPS users. Several ionospheric models for the correction of ionospheric signal delay are available to GPS users. Modelling studies on the ionosphere have gained momentum in recent years, especially after the 1970s. Various important models are described below.

A number of empirical models have been developed to forecast ionospheric delay variability in various regions worldwide; these models include the Bent model, which was developed by Rodney Bent and Sigrid Llewellyn in 1973 [31] to track satellites, ensure satellite intercommunication and correct for the ionospheric delay. In principle, the development of the Bent model involved the fitting of an electron density profile to the database of ionospheric data. This model describes ionospheric electron density as a function of solar radio flux, longitude, latitude, season and time. The Bent model can describe electron density at a given date, location and time. Its resulting profile is classified into five segments: a bi-parabola for the modelling of the

© The Author(s), under exclusive license to Springer Nature Singapore Pte Ltd. 2021
N. A. Elmunim and M. Abdullah, *Ionospheric Delay Investigation and Forecasting*,
SpringerBriefs in Applied Sciences and Technology,
https://doi.org/10.1007/978-981-16-5045-1_4

lower ionosphere, a parabola for the joining of the top-and bottom-side ionospheres and three exponential profile segments, which are combined for the modelling of the top-side ionosphere [35]. However, the Bent model does not involve the lower layers D, E and F1.

The parameterised ionospheric model (PIM) is a climatological global ionospheric and plasmaspheric model that is modelled in accordance with the parameterised output of a number of regional ionospheric models and an empirical plasmaspheric model [16]. It entails four diverse physical models: a low-latitude F layer model, a mid-latitude F layer model, a combination of the low and middle latitude E layer models and a high-latitude E and F layer model. These models are in accordance with a tilted dipole expression of the geomagnetic field and the respective geomagnetic coordinate system [25]. The PIM model describes the worldwide ionospheric amplitude and period and can eliminate only approximately 50% of the ionospheric delay error in real-time [47].

The broadcast ionospheric Klobuchar model [24] was developed by John Klobuchar at the Air Force Geophysics Laboratory, United States. The model algorithm is applied for the correction of the ionospheric delay in GPS for single-frequency communication. This model was developed in 1975. Its algorithm is swift and has minimal complications. One of the major procedures in the design of the algorithm is to obtain the best fit of the daily period with the maximum value of the TEC. The model assumes the ideal smooth behaviour of the ionosphere, and daily substantial alterations cannot be modelled properly. However, the computation memory and capacity of this model are limited. Hence, this model can remove only approximately 50–60% of the ionospheric delay error. Therefore, its accuracy, particularly in the case of high solar activity, is rarely satisfactory even for absolute positioning.

The International Reference Ionosphere (IRI) model is one of the most popularly utilised models and was proposed for international application by the Committee on Space Research (COSPAR), and the International Union Of Radio Science (URSI). It is mainly used to specify ionospheric parameters. IRI describes the monthly averages of the electron density, ion composition, electron temperature, ion temperature and ion drift for a specific date, time and location at altitudes ranging from 60 to 1500 km. The development of this model remains in progress. The IRI's determining and predicting capability of ionospheric behaviours is continuously being updated by the scientific community [6]. This model has been continuously enhanced since its IRI-78 version. [18] reported that the IRI-95 model of TEC overestimates the daily observed minimum TEC values but underestimates other TEC values. [2] utilised the IRI-2000 model for a comparative study of low-latitude GPS-TEC measurements and showed that the modelled TEC is overestimated relative to the observed TEC. The skeleton profile approach [9] has demonstrated that the IRI-2000 model includes a top-side specification. The disadvantage of this method is that the modelled and observed values of the TEC at the high parts of the electron density profile starting at approximately 800 km show numerous inconsistencies because density becomes almost constant at these altitudes.

In general, the earlier IRI models have concentrated on middle-range latitudes successfully, and several researchers that have modelled the TEC over low- and midlatitudes obtained overrated observations due to errors [2]. The probability of spread F occurrence, a feature of observation that takes place near the equator at night, was added to the IRI-2007 version [5] for the first time to enable obtaining a good correlation between the modelled and measured data for low-latitude and equatorial regions. The release of the IRI-2007 model came with updates and a number of improved features that provided good agreement with the measured GPS-TEC for the equatorial region. The performance of the IRI-2007 model was proven during the solar minimum in 2008 by using the STEC data observed from the COSMIC satellite [54]. The IRI model has been steadily enhanced with new data and improved modelling techniques. The current version of IRI is IRI-2012, which was released with several modifications [8]. Relating the model-predicted TEC data and the observed data from the maximum number of stations is crucial because the change in value between the modelled and observed TEC data varies with the location and local time. The IRI-2012 model has three options for top-side electron density: IRI-2001, IRI01-corr and NeQuick. Various top-side electron density options are expressed by various mathematical functions. For example, NeQuick is expressed by the semi-Epstein layer function in terms of height, which depends on the thickness parameter [14]. The IRI-2001 model is expressed by the skeleton profile that contains piecewise constant gradients. For the top-side ionosphere, the electron density profile with the top-side options decreases exponentially with altitude. The electron density profile with the NeQuick and IRI01-corr top-side options decreases with altitude more swiftly than that with the IRI-2001 top-side option [4]. Many researchers worldwide have compared different top-side electron density options to identify the best option for a certain region. The IRI-2016 model is the latest update of the IRI model [3]. A number of researchers have recently investigated the performance of IRI-2016 over high and low latitudes [32, 49, 52, 53].

The NeQuick is a quasi-experimental model that provides a fast estimation of electron density. It was formulated at the Astronomy and Radio Propagation Laboratory (ARPL) of the Abdus Salam International Centre for Theoretical Physics (ICTP), Trieste, Italy, and the Institute for Geophysics Astrophysics and Meteorology (IGAM) of the University of Graz, Austria. For the expected place and time, the NeQuick model has the capability to provide the electron concentration distribution on the top and bottom of the ionosphere. It also provides a real-time ionospheric correction model for the estimation of STEC quantity from the ground-to-satellite or satellite-to-satellite path for the European Space Agency's (ESA) Galileo users [40].

The neural network (NN) technique consists of interconnected groups of 'neurons', which are used to study the behavioural patterns of nonlinear parameters through the storage of information in interconnections and for the subsequent generalisations of the variations in the parameter under consideration [51]. NNs predict ionospheric parameters by providing a history of the physical processes that affect the ionosphere. Many researchers have utilised NNs to predict ionospheric parameters during different seasons, solar and magnetic activity periods, as well as at different latitudes have utilised NNs to predict ionospheric parameters during different seasons, solar and magnetic activity periods, as well as at different latitudes [46]. However, the NN technique is complex and needs extensive data sampling.

Currently, statistical time-series models are used for ionospheric forecasting. A time-series can be described as a collection of observations with even spacing in time and are measured successively. The observation of GPS data, in this case, the double-difference ionosphere between a reference and a roving satellite at a specific baseline, is a strong example of such a series. Time-series are analysed to understand the primary structure and function that created the data. On the basis of the understanding of these mechanisms, a time-series can be mathematically modelled to predict future observations. Several researchers have successfully developed a series of high-precision ionospheric prediction models, such as the time-series model [41]. The time-series model has numerous advantages over other models: it uses less sampling data, includes a complete modelling theory, has a simple calculation process and good extensionality and delivers high-precision short-term ionospheric predictions [28]. Since Liu et al. [30] proposed a method for predicting the TEC of the global ionosphere by using a time-series model based on IGS ionosphere sample data in 2007, a small number of scholars have also used time-series models to analyse and forecast ionospheric delay. However, most of these researchers forecasted ionospheric delay values during geomagnetically quiet periods, and few have forecasted ionospheric delay values during geomagnetically disturbed periods.

One of these time-series techniques is named the Holt–Winter method. The Holt–Winter method is a statistical time-series approach that is used to estimate a future value to generate repeated trends and seasonal time-series patterns [48]. The exponential smoothing function method is applied to reduce variations in time-series data to provide a clear view of the time-series. Moreover, it has the capability to forecast the future values of the time-series data by providing the best approach to obtain measurements with high accuracy. Suwantragul studied the application of the Holt–Winter model in forecasting the ionospheric delay over Chiang Mai in Thailand by using GPS-TEC measurements [48]. Five-day forecasts were generated by using the basic values of the parameters in the Holt–Winter equations and compared with the observed data. The method was developed to predict ionospheric delays. It enhanced positional accuracy by up to 50%. However, further research on modelling with the Holt–Winter model by using various parameters to enhance the accuracy of the method does not exist.

The Autoregressive Integrated, Moving Average (ARIMA) method is a time-series technique that was created by Box and Jenkins in 1976. This technique allows the production of a set of weighted coefficients that describe the ionosphere's behaviour or rate of change during a sample period. These coefficients can then be used to forecast future observations. The ARIMA model takes historical data and decomposes that data into an autoregressive (AR) process, which retains the memory of past events, via an integrated (I) process that renders the data stationary for easy forecasting and generates a moving average (MA) error during forecasting. It does not suffer from the existence of a serial correlation between the error residuals and their own lagged values. Generally, the model is known as ARIMA (p, d, q), where

AR is designated as p, which represents the lingering effects of previous observations. The integrated I component is designated as d, which represents the degree of variation involved. The MA component is designated as q. A number of models have been formulated and proposed to forecast and generate time-series data based on ARIMA processes. A homogeneous nonstationary time-series can be transformed into a stationary time-series by using an appropriate degree of differencing. The ARIMA model is used to model nonstationary events and forms a vital part of the Box–Jenkins method for time-series modelling [43].

4.2 Comparison of Ionospheric Models

The IRI and the Bent models compared with the applications in the determination of satellite orbit [7]. The IRI model gives a better result than the Bent model because it offers a detailed representation of the bottom-side density structure. However, the IRI model shows better accuracy in the prediction of the bottom-side ionosphere than in the prediction of the top-side ionosphere because it is mainly obtained from ionosonde data. Additionally, Okoh et al. [38] used the NeQuick model as the top-side option for modelling over Nigeria to overcome this situation. Meza et al. [35] compared the VTEC values from Bent, IRI and GPS. They chose a number of geomagnetically quiet days because the Bent and IRI models were developed to work under these conditions. Their resu lts showed that amongst the tested models, the GPS model had the best global VTEC representation at any latitude and longitude.

Stankov et al. [47] compared the top-side ionosphere NeQuick and PIM models. They analysed the results of their comparison and provided suggestions to further improve the models. Venkatesh et al. [50] compared the performance of IRI-2012 with that of the NeQuick2 model during the increasing phase of the solar cycle in 2010–2013. They revealed several differences between the measured TEC and the TEC modelled by the IRI-2012 and NeQuick2 models. Prasad et al. [39] compared TEC estimation by GPS-TEC and IRI-2007 by using three different top-side electron density options (IRI-2001, IRI101-corr and NeQuick). Adewale et al. [1] grouped the values of VTEC into four seasons as follows: March equinox (February, March and April), June solstice (May, June and July), September equinox (August, September and October) and December Solstice (November, December and January). In both studies, the TEC obtained with the NeQuick and IRI01-corr options had a good agreement with the GPS-TEC data, whereas the TEC obtained from the IRI-2001 model largely deviated from the GPS-TEC data. Chauhan et al. [12] reported that the GPS-TEC daytime data were in close agreement with the NeQuick and the IRI01-corr data and that the night-time values generated by the models correlated well with those provided by the IRI-2001 model. By contrast, Sethi et al. [45] revealed that the IRI01-corr had better agreement with the observed TEC during the daytime, whereas the NeQuick model showed a better agreement during the night-time. Adewale et al. [1] revealed good agreement between the measured TEC and NeQuick model predictions for the night-time and daytime over the equatorial African sector.

The data used by Chauhan et al. [12] were collected from low- and mid-latitudes stations in the Indian sector, and the data used by Sethi et al. [45] were collected from only one equatorial station from the Indian sector. By contrast, the data used by Adewale et al. [1] were obtained from 13 GPS stations in the African region.

Leong et al. [27] compared the performances of the NeQuick, IRI-2012 and IRI-2007 models at Banting, Malaysia. They also compared the modelled VTEC values with the measured VTEC values collected from dual-frequency GPS data for the year 2011. They found that the seasonal, semiannual and annual variations of equatorial TEC values showed the minimum value during solstice months and the maximum values during equinoctial months. In general, the ionospheric TEC generated by the IRI and the NeQuick models was in good agreement with the GPS-TEC VTEC. For diurnal variation, the NeQuick and the IRI-2007 models exhibited good agreement during the postmidnight and post noon periods, respectively. However, the IRI-2012 model provided a better result for monthly variation than the NeQuick and the IRI-2007 models. Mengistu et al. [34] evaluated the performance of IRI-2016 in the equatorial East African sector. The IRI-2016 model showed better agreement with observations obtained during the solar minimum than with observations taken during other phases of solar activity.

Habarulema et al. [21] compared the NN model with the IRI-2001 model and proved that the IRI-2001 model provided a more accurate prediction for the spring and equinox seasons in South Africa than the NN model. Additionally, Mallika et al. [32] investigated the performance of the NN model and compared it with that of the IRI-2012 and IRI-2016 models over low latitudes. This kind of comparison is of vital importance for the improvement of the IRI model and other TEC models.

References

1. A.O. Adewale, E.O. Oyeyemi, P.J. Cilliers, L.A. McKinnell, A.B. Adeloye, Low solar activity variability and IRI 2007 predictability of equatorial Africa GPS TEC. Adv. Space Res. **49**(2), 316–326 (2012). https://doi.org/10.1016/j.asr.2011.09.032
2. P. Bhuyan, R. Borah, TEC derived from GPS network in India and comparison with the IRI. Adv. Space Res. (2007). http://www.sciencedirect.com/science/article/pii/S027311770700083X
3. D. Bilitza, D. Altadill, V. Truhlik, V. Shubin, I. Galkin, B. Reinisch, X. Huang, International Reference Ionosphere 2016: From ionospheric climate to real-time weather predictions. Space Weather **15**(2), 418–429 (2017). https://doi.org/10.1002/2016SW001593
4. D. Bilitza, International reference ionosphere 2000. Radio Sci. **36**(2), 261–275 (2001). http://onlinelibrary.wiley.com/doi/https://doi.org/10.1029/2000RS002432/full
5. D. Bilitza, International reference ionosphere 2000: examples of improvements and new features. Adv. Space Res. **31**(3) (2003). http://www.sciencedirect.com/science/article/pii/S0273117708000288
6. D. Bilitza, Evaluation of the IRI-2007 model options for the topside electron density. Adv. Space Res. **44**(6), 701–706 (2009). https://doi.org/10.1016/j.asr.2009.04.036
7. D. Bilitza, K. Rawer, S. Pallaschke, Study of ionospheric models for satellite orbit determination. Radio Sci. **23**(3), 223–232 (1988). https://doi.org/10.1029/RS023i003p00223

8. D. Bilitza, D. Altadill, Y. Zhang, C. Mertens, V. Truhlik, P. Richards, L.-A. McKinnell, B. Reinisch, The International Reference Ionosphere 2012 – a model of international collaboration. J. Space Weather Space Clim. **4**, A07 (2014). https://doi.org/10.1051/swsc/2014004

9. H.G. Booker, Fitting of multi-region ionospheric profiles of electron density by a single analytic function of height. J. Atmos. Terr. Phys. **39**(5), 619–623 (1977). https://doi.org/10.1016/0021-9169(77)90072-1

10. D. Buresova, L.R. Cander, A. Vernon, B. Zolesi, Effectiveness of the IRI-2001-predicted N(h) profile updating with real-time measurements under intense geomagnetic storm conditions over Europe. Adv. Space Res. **37**(5), 1061–1068 (2006). https://doi.org/10.1016/j.asr.2006.02.002

11. M. Chakraborty, S. Kumar, B.K. De, A. Guha, Latitudinal characteristics of GPS derived ionospheric TEC: a comparative study with IRI 2012 model. Ann. Geophys. **57**(5), 1–13 (2014). https://doi.org/10.4401/ag-6438

12. V. Chauhan, O.P. Singh, A morphological study of GPS-TEC data at Agra and their comparison with the IRI model. Adv. Space Res. **46**(3), 280–290 (2010). https://doi.org/10.1016/j.asr.2010.03.018

13. V.R. Chowdhary, N.K. Tripathi, S. Arunpold, D.K. Raju, Variations of total electron content in the equatorial anomaly region in Thailand. Adv. Space Res. **55**(1), 231–242 (2015). https://doi.org/10.1016/j.asr.2014.09.024

14. P. Coïsson, S.M. Radicella, R. Leitinger, B. Nava, Topside electron density in IRI and NeQuick: features and LIMITATIONS. Adv. Space Res. **37**(5), 937–942 (2006). https://doi.org/10.1016/j.asr.2005.09.015

15. P.B. Collection, The accuracy of extrapolation (time series) methods: results of a forecasting competition: ABS. Of Forecasting (1986). http://onlinelibrary.wiley.com/doi/https://doi.org/10.1002/for.3980010202/full

16. R.E. Daniell, L.D. Brown, D.N. Anderson, M.W. Fox, P.H. Doherty, D.T. Decker, J.J. Sojka, R.W. Schunk, Parameterized ionospheric model: a global ionospheric parameterization based on first principles models. Radio Sci. **30**(5), 1499–1510 (1995). https://doi.org/10.1029/95RS01826

17. H. Erdoğan, N. Arslan, Identification of vertical total electron content by time series analysis. Digit. Sig. Process. **19**, 740–749 (2009). https://doi.org/10.1016/j.dsp.2008.07.002

18. R.G. Ezquer, C.A. Jadur, M. Mosert de Gonzalez, IRI-95 tec predictions for the south american peak of the equatorial anomaly. Adv. Space Res. **22**(6), 811–814 (1998). https://doi.org/10.1016/S0273-1177(98)00103-3

19. R.G. Ezquer, GPS—VTEC measurements and IRI predictions in the South American sector. Adv. Space **34**, 2035–2043 (2004). https://doi.org/10.1016/j.asr.2004.03.015

20. D.L. Gallagher, P.D. Craven, R.H. Comfort, N. Marshall, S. Flight, An empirical model of the earth' s plasmasphere. Adv. Space Res. **8**(8) (1988)

21. J.B. Habarulema, L.A. McKinnell, P.J. Cilliers, B.D.L. Opperman, Application of neural networks to South African GPS TEC modelling. Adv. Space Res. **43**(11), 1711–1720 (2009). https://doi.org/10.1016/j.asr.2008.08.020

22. S.D.O. Ilboudo, Sombi??, I., Soubeiga, A. K., & Dr??bel, T. , Facteurs influen??ant le refus de consulter au centre de sant?? dans la r??gion rurale Ouest du Burkina Faso. Sante Publique **28**(3), 391–397 (2016). https://doi.org/10.1017/CBO9781107415324.004

23. Y. Kakinami, J.Y. Liu, L.C. Tsai, A comparison of a model using the FORMOSAT-3/COSMIC data with the IRI model. Earth Planets Space **64**(6), 545–551 (2012). https://doi.org/10.5047/eps.2011.10.017

24. J.A. Klobucher, Ionospheric Effects on GPS, Global Positioning System: Theory and Applications, in *Fundamentals of Signal Tracking Theory 163*. ed. by B.W. Parkinson, J.J. Spilker, P. Axelrad, P. Enge (American Institute of Aeronautics and Astronautics, 1996), pp. 485–515

25. A. Komjathi, *Global Ionospheric Total Electron Content Mapping Using the Global Positioning System*. University of New Brunswick, 1997

26. S. Kumar, E. Tan, S. Razul, C.M. See, D. Siingh, Validation of the IRI-2012 model with GPS-based ground observation over a low-latitude Singapore station. Earth, Planets and Space **66**(1), 17 (2014). https://doi.org/10.1186/1880-5981-66-17

27. S.K. Leong, T.A. Musa, K. Omar, M.D. Subari, N.B. Pathy, M.F. Asillam, Assessment of ionosphere models at Banting: Performance of IRI-2007, IRI-2012 and NeQuick 2 models during the ascending phase of Solar Cycle 24. Adv. Space Res. **55**(8), 1928–1940 (2015). https://doi.org/10.1016/j.asr.2014.01.026

28. X.-H. Li, D.-Z. Guo, Prediction of ionospheric Total Electron Content based on semiparametric autoregressive model. Scii. Surv. Map. **36**(2), 149–151 (2011)

29. Z. Li, Z. Cheng, C. Feng, W. Li, A study of prediction models for ionosphere. J. Geophys. **50**(2), 307–319 (2007). http://onlinelibrary.wiley.com/doi/https://doi.org/10.1002/cjg2.1038/full

30. X. Liu, L. Song, X. Yang, G. Yang, Predicting shortdated ionospheric TEC based on wavelet neural network. Hydrograph. Surv. **30**(5), 49–51 (2010). http://en.cnki.com.cn/Article_en/CJF DTOTAL-HYCH201005017.htm

31. S. Llewellyn, R. Bent, Documentation and description of the Bent ionospheric model. In *Technical Report AFCRL-TR-73–0657* (Issue July 1973) (1973). http://oai.dtic.mil/oai/oai?verb= getRecord&metadataPrefix=html&identifier=AD0772733

32. L. Mallika, Ratnam, D.V. Raman, S. Sivavaraprasad, G. Ratnam, B.D.V., D. In, Performance analysis of neural networks with IRI-2016 and IRI-2012 models over Indian low-latitude GPS stations. Astrophys. Space Sci. **365**, 124 (2020) . https://doi.org/10.1007/s10509-020-03821-6

33. L.A. McKinnell, E.O. Oyeyemi, Progress towards a new global foF2 model for the International Reference Ionosphere (IRI). Adv. Space Res. **43**(11), 1770–1775 (2009). https://doi.org/10. 1016/j.asr.2008.09.035

34. E. Mengistu, M.B. Moldwin, B. Damtie, M. Nigussie, The performance of IRI-2016 in the African sector of equatorial ionosphere for different geomagnetic conditions and time scales. J. Atmos. Solar Terr. Phys. **186**, 116–138 (2019). https://doi.org/10.1016/j.jastp.2019.02.006

35. A.M. Meza, C.A. Brunini, W. Bosch, M.A. VanZele, Comparing vertical total electron content from GPS, Bent and IRI models with TOPEX-Poseidon. Adv. Space Res. **30**(2), 401–406 (2002). https://doi.org/10.1016/S0273-1177(02)00314-9

36. M. Mosert, M. Gende, C. Brunini, R. Ezquer, D. Altadill, Comparisons of IRI TEC predictions with GPS and digisonde measurements at Ebro. Adv. Space Res. **39**(5), 841–847 (2007). https:// doi.org/10.1016/j.asr.2006.10.020

37. M. Mosert, S.M. Radicella, D. Buresova, R. Ezquer, C. Jadur, Study of the variations of the electron density at 170 km. Adv. Space Res. **29**(6), 937–941 (2002). https://doi.org/10.1016/ S0273-1177(02)00061-3

38. D. Okoh, L.A. McKinnell, P. Cilliers, P. Okeke, Using GPS-TEC data to calibrate VTEC computed with the IRI model over Nigeria. Adv. Space Res. **52**(10), 1791–1797 (2013). https:// doi.org/10.1016/j.asr.2012.11.013

39. S.N.V.S. Prasad, P.V.S. Rama Rao, D.S.V.V.D. Prasad, K. Venkatesh, K. Niranjan, On the variabilities of the Total Electron Content (TEC) over the Indian low latitude sector. Adv. Space Res. **49**(5), 898–913 (2012). https://doi.org/10.1016/j.asr.2011.12.020

40. S. Radicella, R. Leitinger, The evolution of the DGR approach to model electron density profiles. Adv. Space Res. **27**(1), 35–40 (2001). https://doi.org/10.1016/S0273-1177(00)001 38-1

41. M. Rajabi, A. Amiri-Simkooei, H. Nahavandchi, V. Nafisi, Modeling and prediction of regular ionospheric variations and deterministic anomalies. Remote Sens. **12**(6) (2020). https://doi. org/10.3390/rs12060936

42. V.S. Rathore, S. Kumar, A.K. Singh, A statistical comparison of IRI TEC prediction with GPS TEC measurement over Varanasi, India. J. Atmos. Solar Terr. Phys. **124**, 1–9 (2015). https:// doi.org/10.1016/j.jastp.2015.01.006

43. D. V. Ratnam, B. V. Dinesh, B. Tejaswi, D. P. Kumar, T. V. Ritesh, P. S. Brahmanadam, G. Vindhya, TEC prediction model using neural networks over a low latitude GPS station. Int. J. Soft Comput. Eng. (IJDCE), **2**, 517–521 (2012). http://citeseerx.ist.psu.edu/viewdoc/dow nload?doi=10.1.1.465.1731&rep=rep1&type=pdf

44. L.A. Scidá, R.G. Ezquer, M.A. Cabrera, M. Mosert, IRI 2001/90 TEC predictions over a low latitude station. Adv. Space Res. **44**(6), 736–741 (2009). https://doi.org/10.1016/j.asr.2009. 04.028

45. N.K. Sethi, R.S. Dabas, S.K. Sarkar, Validation of IRI-2007 against TEC observations during low solar activity over Indian sector. J. Atmos. Solar Terr. Phys. **73**(7–8), 751–759 (2011). https://doi.org/10.1016/j.jastp.2011.02.011

46. G. Sivavaraprasad, V.S. Deepika, D. SreenivasaRao, M. Ravi Kumar, M. Sridhar, Performance evaluation of neural network TEC forecasting models over equatorial low-latitude Indian GNSS station. Geodesy Geodynamics **11**(3), 192–201 (2020). https://doi.org/10.1016/j.geog.2019.11.002

47. S.M. Stankov, P. Marinov, I. Kutiev, Comparison of NeQuick, PIM, and TSM model results for the topside ionospheric plasma scale and transition heights. Adv. Space Res. **39**(5), 767–773 (2007). https://doi.org/10.1016/j.asr.2006.10.023

48. S. Suwantragul, P. Rakariyatham, T. Komolmis, A. Sang-In, A modelling of ionospheric delay over Chiang Mai province. IEEE International Symposium on Circuits and Systems, ISCAS 2003, **2**, 340–343 (2003). http://ieeexplore.ieee.org/xpls/abs_all.jsp?arnumber=1205977

49. H.P.T. Thu, C. A. Mazaudier, M. Le Huy, D. N. Thanh, H. Luu Viet, N. Luong Thi, K. Hozumi, T. Le Truong, Comparison between IRI-2012, IRI-2016 models and F2 peak parameters in two stations of the EIA in Vietnam during different solar activity periods. Adv. Space Res. (2020). https://doi.org/10.1016/j.asr.2020.07.017

50. K. Venkatesh, P.V.S. Rama Rao, P.L. Saranya, D.S.V.V.D. Prasad, N. K. , Vertical electron density and topside effective scale height (HT) variations over the Indian equatorial and low latitude stations. Ann Geophys **29**, 1861–1872 (2011)

51. M. Watson, *Common LISP Modules: Artifical Intelligence in the Era of Neural Networks and Chaos Theory*. (Springer, 1991)

52. C. Yang, J. Guo, T. Geng, Q. Zhao, K. Jiang, X. Xie, Y. Lv (n.d.), *Remote sensing Assessment and Comparison of Broadcast Ionospheric Models: NTCM-BC, BDGIM, and Klobuchar*. https://doi.org/10.3390/rs12071215

53. C. Yang, J. Guo, T. Geng, Q. Zhao, K. Jiang, X. Xie, Y. Lv, Assessment and comparison of broadcast ionospheric models: NTCM-BC, BDGIM, and Klobuchar. Remote Sens. **12**(7), 1215 (2020). https://doi.org/10.3390/rs12071215

54. X. Yue, W.S. Schreiner, C. Rocken, Y.H. Kuo, Validate the IRI2007 model by the COSMIC slant TEC data during the extremely solar minimum of 2008. Adv. Space Res. **51**(4), 647–653 (2013). https://doi.org/10.1016/j.asr.2011.08.011

Chapter 5
Ionospheric Delay Forecasting

As discussed in Chap. 3, the ionosphere is a notable source of error that disrupts the accuracy of GPS signals propagating to the ground by changing the speed and direction of signal propagation causing a delay in signals. Therefore, forecasting ionospheric delay is highly important for reducing GPS positioning errors. The statistical Holt–Winter method was chosen due to its suitability for forecasting time-series with repeated trend and seasonal patterns. The Holt–Winter models were compared on the basis of the monthly variations in ionospheric delay. The models were tested by using three different error measurement components to identify the most reliable and suitable model for forecasting ionospheric delay over Malaysia. Moreover, the diurnal, monthly and seasonal variations of the ionospheric delay forecasted for two stations were revealed and validated by using actual ionospheric delay data. Furthermore, the actual and forecasted ionospheric delays during geomagnetic storm activity were investigated. Finally, the ionospheric delay was forecasted by using ARIMA, another statistical time-series model, and the result of the ARIMA was compared with that of the Holt–Winter model. The modified Holt–Winter model provided forecasting results with higher accuracy than the ARIMA model. MAPE was selected to test the accuracy of the investigated method. The method provided good forecasting results considering that its MAPE was less than 20%.

5.1 Holt–Winter Method

The Holt–Winter method is a statistical short-term method that utilises mathematically recursive functions to forecast the behavioural trend of ionospheric delay [15]. It uses a time-series with a repeated trend and a seasonal pattern to forecast under the assumption that the future will follow a similar pattern. The exponential smoothing

© The Author(s), under exclusive license to Springer Nature Singapore Pte Ltd. 2021
N. A. Elmunim and M. Abdullah, *Ionospheric Delay Investigation and Forecasting*,
SpringerBriefs in Applied Sciences and Technology,
https://doi.org/10.1007/978-981-16-5045-1_5

function method is used to minimise the data fluctuations of a time-series to offer a clear perspective of the time-series. Moreover, it can forecast the values of future data in a time-series by giving the best way to derive accurate readings. Three smoothing constants or weight constants, including level (α), trend (β), and seasonal (γ). These smooth constants are suitable for a certain period in updating the component t. The value of the constant in the basic equation is usually 0.2. However, this value can vary between 0 and 1. The Holt–Winter method, when modelled by using the identified parameters, can provide forecasts with increased accuracy [8].

The Holt–Winter forecasting process proceeds as follows: firstly, the data are modified seasonally. Then, forecasts are made for the seasonally controlled data via linear exponential smoothing. Lastly, the forecasts that are modified seasonally are 'reseasonalised' to generate forecasts for the appropriate series. The first step in seasonal modification is the computation of a centred moving average by taking the average of 3 h of data. The ratio is then estimated to a moving average, i.e. the reference data divided by the moving average in each period. The computed seasonal index for each season is determined by estimating the average of all of the ratios for that season and averaging. Subsequently, the ratios are averaged separately for each hour of the day to obtain the non-normalised seasonal indices.

The Holt–Winter method involves two main models in accordance with the type of seasonality: the additive (A-HW) and multiplicative (M-HW) models. The A-HW model is unaffected by changes in data-series and thus works best when the seasonal pattern does not change over time. By contrast, the M-HW model is dependent on data size. For example, ionospheric delay is affected by several factors, such as solar activity. When these factors increase ionospheric delay, the seasonal component of the M-HW model also increases. The A-HW model is applied by using the following equations:

Level:

$$L_t = \alpha(Y_t - S_{t-s}) + (1 - \alpha)(L_{t-1} + b_{t-1}) \tag{5.1}$$

Trend:

$$b_t = \beta(L_t - L_{t-1}) + (1 - \beta)b_{t-1} \tag{5.2}$$

Seasonal:

$$S_t = \gamma(Y_t - L_t) + (1 - \gamma)S_{t-s} \tag{5.3}$$

Fitted:

$$F_t = L_{t-1} + b_{t-1} + S_{t-s} \tag{5.4}$$

Forecast:

$$F_{t+m} = L_t + b_t m + S_{t-s+m} \tag{5.5}$$

The Holt–Winter algorithm requires starting or initialising values. The seasonal component initial value S_1 must be estimated by using Eq. (5.6). The seasonal duration level s is represented by Eq. (5.7) as

$$S_1 = Y_1 - L_s, \; S_2 = Y_2 - L_s, \ldots, \; S_s = Y_s - L_s \tag{5.6}$$

$$L_s = \frac{1}{s}(Y_1 + Y_2 + \cdots + Y_s) \tag{5.7}$$

The equations utilised for the M-HW seasonal model are

Level:

$$L_t = \alpha \frac{Y_t}{S_{t-1}}(Y_t - S_{t-1}) + (1-\alpha)(L_{t-1} + b_{t-1}) \tag{5.8}$$

Trend:

$$b_t = \beta(L_t - L_{t-1}) + (1+\beta)b_{t-1} \tag{5.9}$$

Seasonal:

$$S_t = \gamma \frac{Y_t}{L_t} + (1-\gamma)S_{t-s} \tag{5.10}$$

Fitted:

$$F_t = (L_{t-1} + b_{t-1})S_{t-1} \tag{5.11}$$

Forecast:

$$F_{t+m} = (L_t + b_t m)S_{t-s+m} \tag{5.12}$$

The seasonal component initial value S_1 and the seasonal duration level s are estimated by using the following equations:

$$S_1 = \frac{Y_1}{L_1}, \; S_2 = \frac{Y_2}{L_2}, \ldots, \; S_s = \frac{Y_s}{L_s} \tag{5.13}$$

$$L_s = \frac{1}{s}(Y_1 + Y_2 + \ldots + Y_s) \tag{5.14}$$

where

L_t	level
b_t	trend
S_t	season
Y_t	VTEC
t	time period of the L_t, b_t, S_t and Y_t components
F_t	forecast value of a period ahead
F_{t+m}	monthly forecasting time period
α	level-smoothing coefficients
β	trend-smoothing coefficients
γ	seasonal-smoothing coefficient
m	forecast period
s	seasonal duration
S_1	initial value of the seasonal component.

5.1.1   Modified Holt–Winter Method

The basic Holt–Winter equation does not provide highly accurate forecasts. Therefore, the Holt–Winter method has been modified to obtain highly accurate forecast results. Each of the initial values for the level, trend and seasonal parameters are tested with different values that are applied to the Holt–Winter equations. Ionospheric delay is forecasted and accuracy is calculated at every point in time because the initial values for the level, trend and season (α, β and γ) change independently. The initial values for the level, trend and season that give the best and most accurate forecast results are 0.9, 0.1 and 0.1, respectively. This new equation was developed by using MATLAB programming code [6]. The forecasted results have been validated and tested and have been found to be more accurate than the results obtained from basic equations. The new equations are given as follows:

Level:

$$L_t = 0.9\frac{Y_t}{S_{t-1}}(Y_t - S_{t-1}) + (1 - 0.9)(L_{t-1} + b_{t-1}) \qquad (5.15)$$

Trend:

$$b_t = 0.1(L_t - L_{t-1}) + (1 + 0.1)b_{t-1} \qquad (5.16)$$

Seasonal:

$$S_t = 0.1\frac{Y_t}{L_t} + (1 - 0.1)S_{t-s} \qquad (5.17)$$

Fitted:

$$F_t = (L_{t-1} + b_{t-1})S_{t-1} \tag{5.18}$$

Forecast:

$$F_{t+m} = (L_t + b_t m)S_{t-s+m} \tag{5.19}$$

For forecasting, 24 h data are used, where the median of 1 h interval data taken for each hour for all of the PRN median data is taken. The hourly 3-day (72 h) data is used to forecast the following day.

5.2 Error Measurements

Forecasting models must be evaluated from different perspectives, such as the extent of the deviation of the forecasted values from actual values, the utility of the model in forecasting and the strength of the linear correlation between independent and dependent variables. The forecasted values obtained from various forecasting models may differ. The components of error measurement are effective in measuring the appropriateness and accuracy of a forecasting model and in demonstrating the effectiveness of a forecasting model. A model is tested by taking the difference between the forecasted and actual values: small differences are indicative of good models. Various error measurement criteria, such as the mean absolute percentage error (MAPE), mean absolute division (MAD) and mean square division (MSD), can be used to compare various forecasting models.

The MSD and MAD express accuracy in metres. Such an expression is helpful for conceptualising the amount of error. Outliers have a lesser effect on the MAD than on the MSD. The unit of MAD is similar to that of the time-series expressed in Eq. (5.20). The MSD is commonly estimated by using a similar number of total observations n irrespective of the kind of model in Eq. (5.21). The MAPE is the sum of absolute errors divided by actual values. It gives the value of the error in relation to the observed value i. The MAPE is utilised to define the error between the forecasted results and the actual data and to validate the effectiveness of a model. Lewis [14] interpreted MAPE values to judge forecasting accuracy. The MAPE is applied by using Eq. (5.22), and the percentage error is calculated with Eq. (5.23) below:

$$\text{MAD} = \frac{1}{n}\sum_{t=1}^{n}|Y_t - F_t| \tag{5.20}$$

$$\text{MSD} = \frac{1}{n}\sum_{t=1}^{n}(Y_t - F_t)^2 \tag{5.21}$$

$$\text{MAPE} = \frac{1}{n} \sum_{t=1}^{n} |PE_t| \tag{5.22}$$

$$PE_t = \left(\frac{Y_t - F_t}{Y_t} \right) \times 100 \tag{5.23}$$

where

n number of total observations
Y_t actual value
F_t forecast value of period t
PE_t percentage of the error

MAPE values of less than 10% indicate that a model gives a highly accurate forecast, those between from 10 to 20% indicate that the model provides a good forecast and those between from 20 to 50% indicate a reasonable forecast. A range of greater than 50% indicates that the model yields an inaccurate forecast [14].

5.3 Data Measurements and Forecasting

The dual-frequency GISTM receiver was used to estimate ionospheric TEC data. As shown in Fig. 5.1, GISTM receivers have been installed at two different locations in Malaysia. The first GISTM receiver is located at the Space Science Centre, Institute of Climate Change, UKM. The antenna is placed on the roof of the faculty of engineering and built environment at the geographic coordinates of 2.55°N–101.46°E and the geomagnetic coordinates of 7.10°S–174.05°E. The second GISTM receiver is located at Langkawi at the geographic coordinates of 6.19°N–99.51°E and geomagnetic coordinates of 3.39°S–172.42°E.

5.4 GISTM

The GPS Silicon Valley's GISTM system (model GSV4004B) can track up to 11 GPS signals at the L1 and L2 frequencies of 1575.42 and 1227.60 MHz, respectively [19], and is designed specifically to record the ionospheric TEC on the L1/L2 frequencies. The GISTM GSV4004B consists of three main components: a GPS receiver (NovAtel's EuroPak-3 M), a DC power supply with different interconnecting cables and a NovAtel GPS-533 L1/L2 GPS 702 antenna. As presented in Fig. 5.2, the NovAtel GPS-533 L1/L2 GPS 702 antenna is connected to the GSV4004B unit that is interfaced with a PC to continuously acquire and store data.

The receiver of the GISTM model GSV4004B incorporates software for spatial data collection in a personal computer. In addition to its core capability to log all

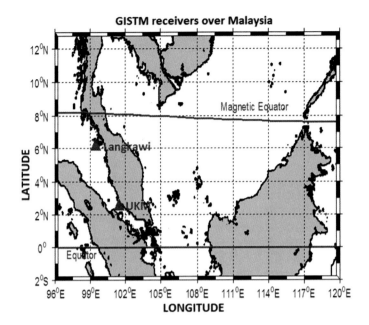

Fig. 5.1 GISTM station networks

Fig. 5.2 Schematic of the GISTM GSV4004B receiver

special ionospheric data, the software permits the real-time study of TEC specifications, as well as other relevant data for all satellites. Given that the collection of raw 50 Hz data comes with excessive storage requirements, thresholds can be assigned to specific parameters to remove redundant data in real-time. Thus, this GISTM system provides the real amplitude, single-frequency carrier-phase measured data and dual-frequency TEC measured data from 8–11 GPS satellites in view. The main aim of the GISTM GSV400B is to gather the TEC data for all known GPS satellites that have up to three SBAS-GEO satellites.

5.4.1 GISTM Data Recording

Once the GSV4004B system is powered, it initialises automatically and starts to track satellites by using default information. The clock time is adjusted, and the receiver bias is entered to set the location. However, to ensure that accurate TEC values are obtained from the GSV4004B, the bias is entered into the GISTM GSV4004B receiver. This receiver bias is supplied by the manufacturer. For satellite biases, the values of the 32 obtained offsets are processed as the initialisation of the receiver proceeds by using the CP-Offset command.

As given by Eq. (5.24), the TEC is measured by the number of electrons through the path of the signal from the satellite to the receiver in a 1 m^2 cylindrical shell.

$$STEC = \int_r^s N_e dI \tag{5.24}$$

where

N_e electron density along the LOS dl

The measurements are influenced by the elevation angle of the satellite from the station and ionospheric height [12]. A TEC unit is written as 1×10^{16} electrons m^2. The TEC determined by GISTM is termed as the STEC and is dependent on the ray path geometry via the ionosphere. The STEC is proportional to the ionospheric delay between the L1 and L2 signals. The SLOG data-recording program utilises this delay and addresses instrumental biases prior to the determination of the final STEC values by using the formula from [19] as follows:

$$STEC = \left[9.483 * \left(PR_{L2} - PR_{L1} - \Delta_{C/A-P,PRN}\right)\right.$$
$$\left. + TEC_{RX} + TEC_{CAL}\right]TECU \tag{5.25}$$

where

PR_{L2} L2 pseudorange in metres.
PR_{L1} L1 pseudorange in metres.

$\Delta_{C/A\text{-}P,PRN}$ input bias between SV C/A- and P-code chip transition in metres.

TEC_{RX} TEC result due to internal receiver L1/L2 delay.

TEC_{CAL} user-defined TEC offset, which is 50, used for TEC calibration as recommended in the user manual.

The SLOG program is used to log the data from the receiver as mentioned earlier. SLOG is an application based on Microsoft Windows that is implemented via the command prompt. SLOG is programmed to create a new file on a daily basis and to store these files in a folder named in accordance with the GPS week. After each GPS week, a new folder is automatically produced by the program itself, and the corresponding daily files are then stored in it. The GPS days start from Sunday (0) to Saturday (6). The final STEC data are stored in SLOG files within the GPS format (a binary format) and then converted into ASCII format by using the Parselsmr.exe program. The Parseismr.exe program is accessed via the command prompt and firstly converts the GPS file format (.gps) into text formats (.txt) and subsequently into Microsoft Excel format (.xls).

5.4.2 GISTM Data Processing

A 50 Hz measurement is used at the amplitude and phase, and a 1 Hz measurement is applied for the divergence of the satellite's code/carrier. The GPS data are analysed by using an elevation angle greater than 20° because the GPS signal suffers from considerable fluctuations due to the multipath effect [4] that appears with a small elevation angle. The lock time shows the amount of time the receiver is restricted to the carrier phase on the L1 and L2 (L1, L2 lock time), and the measurement data for the L1 and L2 are restricted for over 240 s (4 min). This time is required by the GISTM receiver to enable the detrending high-pass filter to reinitialise when the lock to the carrier phase is lost. The application of a dual-frequency pseudorange and carrier phase allows the calculation of the STEC through the LOS. The GISTM receiver provides the STEC data at an interval of 60 s. These data are transformed into the equivalent VTEC by dividing them by the secant of the elevation angle at an average ionospheric height. As explained in Sect. (3.8), VTEC is regarded as a compact parameter that characterises the TEC over the receiver position via the use of the obliquity factor to determine the error from ionospheric delay equations.

5.5 Holt–Winter Models Forecasting

The statistical method is applied to the time-series of GPS-TEC to forecast the ionospheric delay during the period of October 2009 to December 2010 by using the A-HW and M-HW Holt–Winter models. The GISTM receiver over UKM and Langkawi stations is used. The ionospheric delay is forecasted from morning to noon time (08:00–12:00 LT) and from afternoon to night-time (15:00–21:00 LT). These time periods are chosen to indicate temporal variations, such as the ascending and declining phases of ionospheric delay, in ionospheric delay to identify the suitable forecasting model.

5.6 Comparison of the A-HW and M-HW Models

The monthly variations of the actual and forecasted ionospheric delay during 08:00–21:00 LT obtained by using the A-HW and M-HW models are shown in Fig. 5.3 to compare the forecasting results of the A-HW and M-HW models. The figure presents the ionospheric delay in metre units on the vertical axis and the hourly local time on the horizontal axis. Malaysia's local time is 8 h ahead of the universal time (UT). The ionospheric delay forecasted by both models showed diurnal variations with peaks during post noon and rapid decrements at night. This behaviour will be discussed in detail later on. The mean ionospheric delay varied within the range of 1.1–4.6 m. Ionospheric delay increased during March, April, September and November 2010 with a slight difference between the curves of the actual and forecasted delays. In general, the results of the models showed good agreement with the actual ionospheric delay for all months, except for December 2009 and May 2010. However, the A-HW model underestimated the delay in December 2009 by approximately 0.3 m and overestimated the delay in May 2010 by approximately 0.5 m. The overall average of errors between the actual and forecasted delays throughout the considered period fell in the range of 0–2 m.

5.7 Error Variations of the Models

The monthly MAPE, MAD and MSD error measurements were calculated to compare the forecasting performances of the A-HW and M-HW models. These error measurement components were used to describe the errors of the forecasting models. Figure 5.4 depicts the monthly variations in the MAPE, MAD and MSD error measurement components of the A-HW and M-HW models for morning to noon time (08:00–12:00 LT). In the figure, the MAPE is shown in percentages in the top panel, the MAD is presented in metre units in the second panel, the MSD is provided in metre units on the horizontal axis in the bottom panel and the months from October 2009 to December 2010 are given on the vertical axis. As illustrated in Fig. 5.4, the MAPE values of the A-HW model fell in the range of 0.2%–8.2% and those of the M-HW model fell in the range of 0.3%–4.2%. The maximum MAD errors of the A-HW and M-HW models were 0.2 and 0.1 m, respectively. The maximum MSD errors of the A-HW and M-HW models were 0.03 and 0.01 m, respectively. The peaks of the error of the A-HW model were observed during October 2009 and May

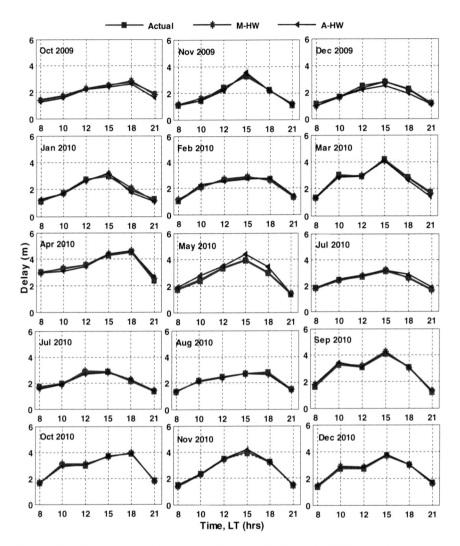

Fig. 5.3 Monthly variations of the actual and forecasted (A-HW and M-HW) ionospheric delay

and July 2010, whereas those of the M-HW model were observed during May and June 2010.

The monthly variations of the MAPE, MAD and MSD error measurements of the A-HW and M-HW models for the afternoon to night-time (15:00–21:00 LT) are shown in Fig. 5.5. The MAPE values of the A-HW and M-HW models fell in the ranges of 1.5%–10.2% and 0.4%–4.6%, respectively. The maximum MAD values of the A-HW and M-HW models were 0.2 and 0.1 m, respectively. The A-HW and M-HW models provided the maximum MSD values of 0.08 and 0.02 m, respectively. The MSD values had the largest difference between the A-HW and M-HW error

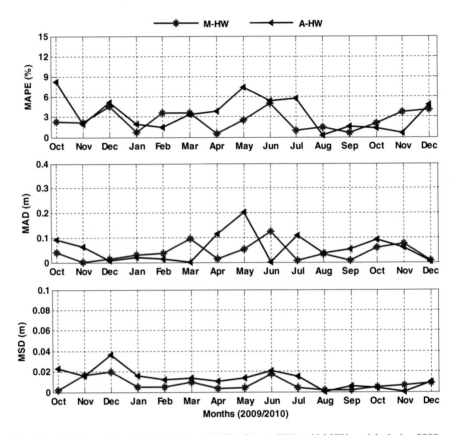

Fig. 5.4 Variations of the MAPE, MAD and MSD of the A-HW and M-HW models during 0800–1200 LT

trends. The MAPE values provided by the A-HW model peaked in October and December 2009 and March, May and September 2010, whereas those given by the M-HW model peaked in January, April and June 2010. Therefore, in both considered periods, the A-HW model exhibited a higher amount of error than the M-HW model.

As explained in Table 5.1, the percentage of the overall average of forecasted error obtained from October 2009 to December 2010 showed that the A-HW model had slightly higher error than the M-HW model.

As inferred from the obtained results, ionospheric delay errors for the afternoon to the night-time period were generally higher than those for the morning to the noon period and ionospheric delay normally peaked in the post noon period and decreased rapidly at night. This behaviour was expected because of the effects of ultraviolet radiation. The ionosphere is a layer that is ionised by the sun's radiation. Given that ionisation is mostly due to extreme ultraviolet radiation, it increases ionospheric delay during the first hours of the afternoon. The forecasting results of both Holt–Winter models showed that the A-HW model had a comparatively higher value of

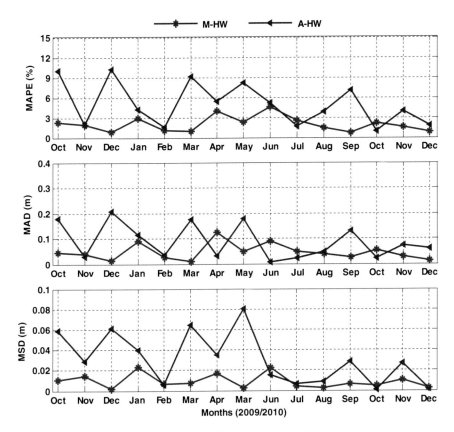

Fig. 5.5 Variations of the MAPE, MAD and MSD values of the A-HW and M-HW models during 15:00–21:00 LT

Table 5.1 Percentage of the overall average of forecasted error

Time	A-HW (%)	M-HW (%)
08:00–12:00 LT	3.5	2.2
15:00–21:00 LT	5	2

the forecast error than the M-HW model [7]. This difference made sense in terms of using one of the Holt–Winter models that can provide superior forecasting results, which are essential for correcting GPS positioning. As could be deduced through the comparative analysis of both models, compared with the A-HW model, the M-HW model was more reliable for forecasting ionospheric delay given its slight error. Hence, only the M-HW model was used in this book.

5.8   Variations in Ionospheric Delay Forecasting

The M-HW model was used to forecast ionospheric delay accurately, to investigate the variations in ionospheric delay and to validate the Holt–Winter model. The diurnal, monthly and seasonal ionospheric delay variations over the UKM station (geographical coordinates: 2.55°N–101.46°E, geomagnetic coordinates: 7.10°S–174.05°E) and Langkawi station (geographical coordinates: 6.19°N–99.51°E, geomagnetic coordinates: 3.39°S–172.42°E) in Malaysia during the year 2011 were estimated from GISTM receivers and compared with the ionospheric delay forecasted by using the Holt–Winter model. Moreover, the differences in the ionospheric delay observed during geomagnetic storm disturbances were investigated.

5.8.1   Diurnal Variations

In equatorial regions, diurnal variations on quiet days normally vary in accordance with photoionisation and the recombination losses associated with local solar radiation. The diurnal variations in ionospheric delay over the UKM and Langkawi stations during a typical quiet day on 12 May 2011, where $Kp \leq 1$, are plotted in Fig. 5.6 to illustrate the comparison between the actual delay (GPS-TEC) and forecasted delay (Holt–Winter) over the UKM and Langkawi stations. The ionospheric delay is displayed on the vertical axis in metres, and the local time is displayed on the horizontal axis in hours. The diurnal pattern of ionospheric delay showed a single obvious peak. It steadily increased from morning to post noon, peaked between 13:00–16:00 LT and then fell to its minimum value during sunrise at 5:00 LT.

Adewale et al. (2011) investigated the diurnal variation in the TEC over several stations in Lagos, Nigeria (geomagnetic coordinates: 8.86°N–76.90°E). They observed peaks from 13:00 to 17:00 LT and the minimum value at 6:00 LT. [1] observed diurnal variations over Bangkok, Thailand (geomagnetic coordinates: 3.76°N–172.90°E) during 2011 and found peaks during the post noon period between 14:00 and 18:00 LT. Therefore, this result indicated that the peaks of ionospheric delay were associated with low chemical losses at high altitudes and the production of solar radiation during the post noon period [11]. The diurnal behaviours of the actual and forecasted ionospheric delays over the UKM and Langkawi stations were similar and were in good agreement with this result. Nevertheless, the result of the Holt–Winter model for ionospheric delay during early morning was overestimated compared with the actual ionospheric delay. The ionospheric delay over the UKM station was slightly higher than that over the Langkawi station. The maximum ionospheric delay over the UKM station reached 5 m, whereas that over the Langkawi station reached 4.9 m. The lower values of ionospheric delay over the UKM and Langkawi stations were 0.8 and 0.6 m, respectively.

Fig. 5.6 Diurnal hourly variations of the actual (GPS-TEC) and forecasted (Holt–Winter) ionospheric delay over the UKM and Langkawi stations

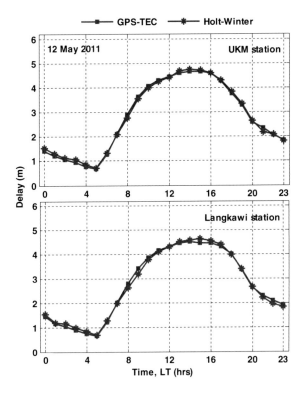

5.8.2 Monthly Variations

The diurnal and monthly medians of ionospheric delay variations over the UKM and Langkawi stations are illustrated in Figs. 5.7 and 5.8. The hourly medians of ionospheric delay were plotted against time for each month to illustrate the monthly variations between the actual and forecasted ionospheric delay. In general, the overall actual and forecasted ionospheric delay trends over the UKM and Langkawi stations showed similar monthly patterns over the course of the year. Ionospheric delays began to steadily increase from February to reach a maximum during March and April and then slightly decreased until September before increasing again in October and November. The monthly variations in the Holt–Winter forecasts showed a good agreement in the GPS-TEC data. The forecast trends were nearly comparable with the actual ionospheric delays at the Langkawi and UKM stations. However, the forecasts for the peak post noon hours showed a slight underestimation of approximately 0.45 m in March, September and November for the UKM station and approximately 0.39 m in March, July and August for the Langkawi station. Compared with the Langkawi station, the UKM station recorded higher ionospheric delays during February, March, June, August, September and October. The obtained results showed that the highest ionospheric delay occurred in the equinoctial months of March,

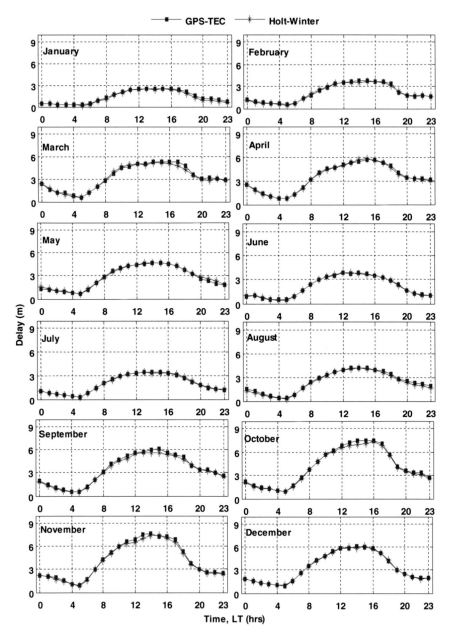

Fig. 5.7 Monthly variations of the actual (GPS-TEC) and forecasted (Holt–Winter) ionospheric delay over the UKM station

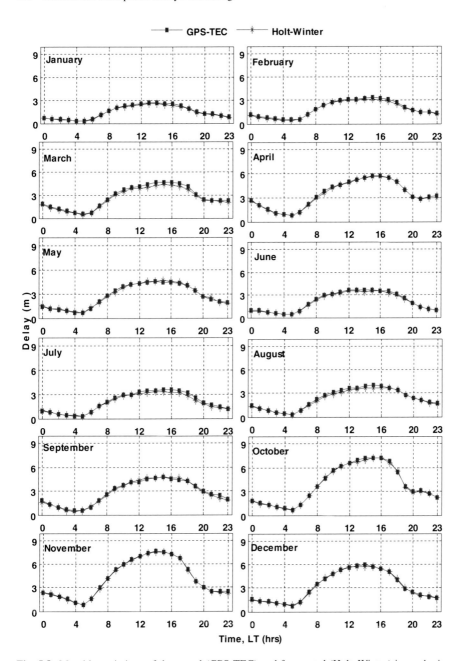

Fig. 5.8 Monthly variations of the actual (GPS-TEC) and forecasted (Holt–Winter) ionospheric delay over the Langkawi station

April, September, October and November, whereas the lowest ionospheric delay occurred in the solstice months of January, June and July. Various researchers have reported similar findings for India (geomagnetic coordinates: 11.84°N–152.82°E), Thailand (geomagnetic coordinates: 3.76°N–172.90°E), Kenya (geomagnetic coordinates: 3.37°S–110.18°E), Uganda (geomagnetic coordinates: 1.14°S–104.85°E) and Ethiopia (geomagnetic coordinates: 5.18°N–114.15°E). They reported that the maximum monthly variations were observed in March and the minimum variations were observed in June and July [1, 13, 16, 17]. The highest monthly diurnal peak for the Langkawi and UKM stations was approximately 8 m and was likely influenced by geomagnetic disturbances in October. The comparable peak during quiet periods was 5 m.

Several moderate geomagnetic storms occurred in 2011. August and September were characterised by a number of mild storms. Submajor geomagnetic storms occurred on 6 August and 26 September with minimum Dst values of approximately − 113 and − 103 nT, respectively. These moderate events did not significantly affect forecast trends. However, the geomagnetic storm that occurred on 25 October had a major effect, the details of which will be described in later sections.

The MAPE measures the errors in the forecasting model. The hourly median of variation in the MAPE and the monthly MAPE averages for the UKM and Langkawi stations are shown in Figs. 5.9 and 5.10, respectively. In the top panel, the hourly percentage of the MAPE is plotted on the vertical axis, and the local time is plotted on the horizontal axis. The bottom panel presents the monthly MAPE on the vertical axis, and the month is presented on the horizontal axis.

As illustrated in the top panels (a) in Figs. 5.9 and 5.10, the maximum MAPE for the UKM and Langkawi stations was observed during morning between the hours of 0:00 LT and 7:00 LT in all the months of 2011. The MAPE value was approximately 8%. As can be seen in Fig. 5.9b, for the UKM station, the maximum monthly average MAPE was observed in March and September with a value of approximately 5%, and the minimum was observed in April and December with the value of 2%. The MAPE's maximum monthly average for the Langkawi station was observed in March and July with a value of approximately 6.2%, whereas the minimum value of approximately 1.6% was observed in April and December as shown in Fig. 5.10b. Furthermore, the Langkawi station had a comparatively higher value of MAPE than the UKM station [6].

5.8.3   Seasonal Variations

The seasons were described as follows to investigate the seasonal variations of ionospheric delay: summer (May, June, July and August), winter (January, February, November and December) and equinox (March, April, September and October). The hourly seasonal variation of ionospheric delay over the UKM and Langkawi stations is given in Fig. 5.11.

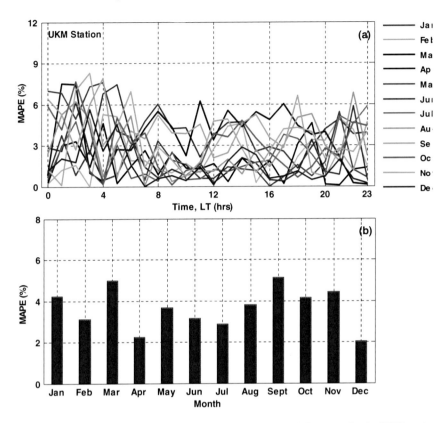

Fig. 5.9 Hourly median of MAPE variations and the monthly MAPE average for the UKM station

The results showed that the maximum value of ionospheric delay was observed during the equinox because the sun shines directly over the equatorial region during the equinoctial months, causing the strongest ionisation over the ionosphere. The minimum ionospheric delay value observed during the summer could be due to the unequal heating of the two hemispheres as a result of the transport of neutral constituents from the summer hemisphere to the winter hemisphere (hot to cold). This phenomenon reduces the recombination rate in winter relative to that the summer and results in the higher electron concentration in the winter than in the summer. The change in the direction of the neutral wind is another possible cause for this seasonal phenomenon. The meridional component of neutral wind blowing from the summer hemisphere to the winter hemisphere can reduce the peak ionisation value during the summer solstice, during which it blows in a direction opposite to the plasma diffusion process originating from the magnetic equator. Consequently, during the equinox, the blowing of meridional winds from the equator to polar regions causes a high ionisation crest value.

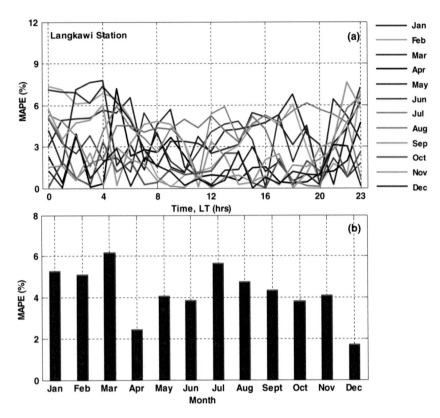

Fig. 5.10 Hourly median of MAPE variations and monthly MAPE average for the Langkawi station

Galav et al. [9] and Kumar et al. [13] studied the seasonal variation in the Indian region and found that the maximum seasonal variation occurred during the equinox, whereas the minimum value was observed in the summer. In this study, the maximum seasonal ionospheric delay was observed at approximately 6 m over the UKM station and at 5 m over the Langkawi station after noontime during the equinox. It then decreased to the minimum of 0.7 and 0.6 m over the UKM and Langkawi stations, respectively. In summer, the maximum ionospheric delay reached 4 m over the UKM station and 3.8 m over the Langkawi station and decreased to a minimum value of 0.4 m over both stations during the early morning. Generally, the ionospheric delay forecasted by the Holt–Winter model exhibited good agreement with the actual GPS-TEC data for all of the seasons and over both stations. However, a slight underestimation of approximately 0.2 m was observed for the Langkawi station during the peak post noon hours of the summer season.

The seasonal variations of the hourly median MAPE for the UKM and Langkawi stations are shown in Fig. 5.12a and b, respectively. The MAPE is presented on the vertical axis, and the hourly local time is given on the horizontal axis. The highest value of the MAPE was observed in the summer. The MAPE values for the UKM

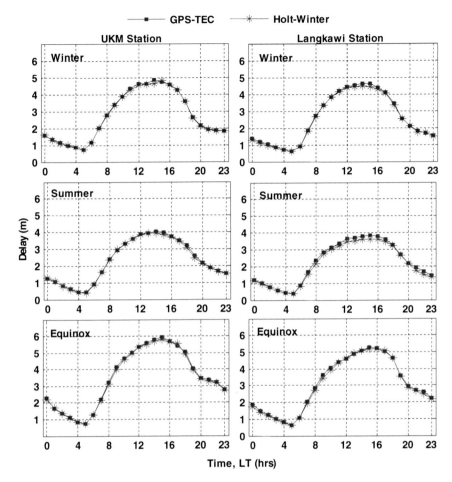

Fig. 5.11 Seasonal variations of the actual (GPS-TEC) and forecasted (Holt–Winter) ionospheric delay over the UKM and Langkawi stations

and Langkawi stations reached 6.2% and 7.9%, respectively. The maximum MAPE values for the UKM and Langkawi stations were observed at 5:00 and 8:00 LT, respectively. The averages of the seasonal MAPE values for the UKM and Langkawi stations are shown in Fig. 5.12c. The vertical axis represents the seasonal average MAPE, and the horizontal axis represents the seasons. The maximum and minimum MAPE values were found in the summer and winter, respectively. The MAPE values for the UKM and Langkawi stations during the summer were 2.4% and 4%, respectively, and those during the winter were 1.7% and 2.3%, respectively. Generally, the MAPE values for the Langkawi station were slightly higher than those for the UKM station.

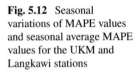

Fig. 5.12 Seasonal variations of MAPE values and seasonal average MAPE values for the UKM and Langkawi stations

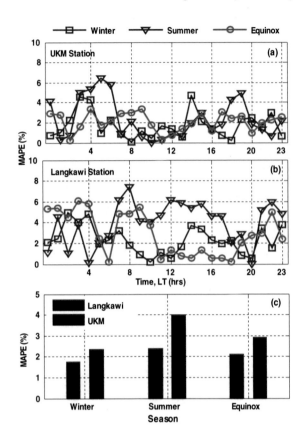

5.9   Perturbation of Ionospheric Delay During Disturbed Periods

The geomagnetic storm that occurred on 25 October 2011 was classified as an intense or major storm. During the storm, the Dst value sharply decreased to − 136 nT and the Kp index reached the maximum value of approximately 7. The diurnal variation of the actual ionospheric delay over the UKM and Langkawi stations during the geomagnetically disturbed days of 23–28 October 2011 is shown in Fig. 5.13. On 25 October 2011, a geomagnetic storm significantly increased the ionospheric delay from 15:00 LT until the next day at 4:00 LT. The peak in the diurnal variation of the actual ionospheric delay was greatly enhanced on 25 October, when it reached 9.3 m over the UKM station and 8.7 m over the Langkawi station. As depicted in Fig. 5.6, the comparison of the diurnal values of the ionospheric delay for the disturbed periods even before and after the geomagnetic storm with those for the quiet day of 12 May revealed that ionospheric delay was lesser during quiet periods than during disturbed periods.

Fig. 5.13 Daily variations of actual ionospheric delay from 23–28 October 2011

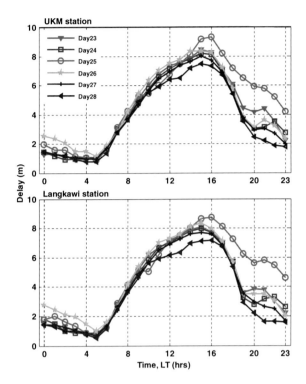

During the disturbance of the geomagnetic storm, energy is input into the polar ionosphere. This event changes several thermospheric parameters, such as composition, temperature and circulation. Changes in composition directly affect the electron concentration in the F2 region and then the TEC, whereas circulation spreads heated gas to low latitudes. [2] reported that during major magnetic storms, the inability of the currents associated with the inner magnetospheric electric field that is directed along the dusk-to-dawn direction to shield the midlatitude and equatorial latitudes from high-latitude electric fields leads to the instantaneous penetration of electric fields from high latitudes to the mid-lattidue and the equatorial ionosphere. Consequently, particle transport and the prompt penetration of the high-latitude electric field into the lower latitude, which travels equatorward with high velocities during the storm, can increase TEC values and ionospheric delay.

The diurnal variation of the ionospheric delay due to the earth's daily rotation during 23–28 October 2011 is shown in Fig. 5.14. The top panel corresponds to the hourly variations of ionospheric delay over the UKM and Langkawi stations; the second panel corresponds to the southward component of the interplanetary magnetic field (IMF) Bz. The next two bottom panels correspond to the two different geomagnetic perturbation indexes Kp and Dst. The Dst and Kp indexes were obtained from the World Data Centre Kyoto, Japan (http://wdc.kugi.kyoto-u.ac.jp/wdc/Sec3.html), and the IMF Bz data were obtained from the OMNI database (http://omniweb.gsfc.nasa.gov/form/dx1.html). On the days of the geomagnetic storms (the dotted line), a

Fig. 5.14 Hourly variations
of ionospheric delay over the
UKM and Langkawi stations
and the parameters of the
IMF Bz component of the
interplanetary magnetic field,
the Kp index and the Dst
index of the geomagnetic
activity

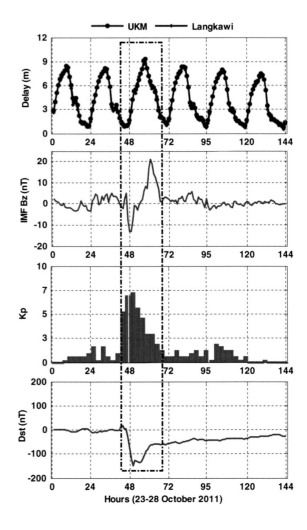

significant increase in the diurnal variations of the ionospheric delay was observed
over both stations. The ionospheric delay reached 9.8 m during the geomagnetic
storm's recovery phase on 25 October. The geomagnetic storm, which was preceded
by a sudden storm commencement event, occurred at 17:00 UT (24 October)/1:00 LT
(25 October). The IMF Bz turned sharply southward at 19:00 UT (24 October)/3:00
LT (25 October) and attained a maximum negative value of − 13 nT at 23:00 UT (24
October)/7:00 LT (25 October) before gradually increasing to its maximum value
of 22 nT at 16:00 UT (25 October)/0:00 LT (26 October). The Kp index started to
increase before the IMF Bz declined. Moreover, it continued to increase even when
the IMF Bz attained its maximum positive value [5]. The Dst started to decrease
from 21:00 UT (25 October)/5:00 LT (24 October) and attained a maximum value
of − 137 nT at 1:00 UT/9:00 LT on 25 October. After the commencement of the

geomagnetic storm, the Dst did not recover to its initial value until 3 days later. At high latitudes, electric field penetration leads to the formation of ionospheric electron density irregularities in the equatorial region. Therefore, the greatest decrease in the Dst value was observed during the sudden intensification of the ring current [2]. The positive value of the Dst could be mainly attributed to the compression of the dayside magnetosphere during the initial phase of the geomagnetic storm, and the negative value was due to magnetic reconnection and ring current formation during the main phase of the geomagnetic storm. Several studies have discussed the perturbations in diurnal variations during geomagnetic storms [2, 18]. Zou et al. [20] used GPS-TEC to observe the same storm on 25 October 2011 and found that ionospheric delay increased at a higher rate during the storm than during the quiet time. This effect was more clearly observed at lower latitudes than at higher latitudes. Other researchers studied the effects of the same geomagnetic storm on 24–25 October 2011 on the ionospheric TEC of two East African stations and found that for both stations, the diurnal variations increased during the storm period [3].

Figure 5.15a shows the variations of the forecasted ionospheric delay over the UKM and Langkawi stations during the period of disturbance from 23 to 29 October 2011. The vertical axis shows the ionospheric delay in metres, and the horizontal axis shows the hourly median of the days of the disturbed period. The variations in the forecasted ionospheric delay for the disturbed period fell in the range of 0.07–11 m for the UKM station and 0.5–10.6 m for the Langkawi station. For both

Fig. 5.15 Variations of ionospheric delay forecasts during disturbed days and MAPE variances during disturbed days for the UKM and Langkawi stations

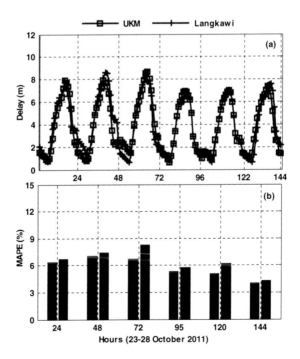

stations, the maximum value for the forecasted ionospheric delay was observed during the geomagnetic storm day (25 October). As shown in Fig. 5.15 (b), the error measurement MAPE was utilised to identify the forecasting model error. The percentage of the MAPE is seen on the vertical axis, and the horizontal axis shows the disturbed days. The plot depicts that the forecasting error for the UKM station was lower than that for the Langkawi station. The maximum forecasting error for the UKM station was 6.6% and that for the Langkawi station was 8.2%. The MAPE value for the disturbed period was slightly higher than that for the quiet period because ionospheric delay increased during the intense geomagnetic storm that occurred on 25 October. This increase affected forecasting and slightly increased forecasting error [6]. The accuracies of the method during the quiet and disturbed periods were 98% and 93%, respectively.

However, ionospheric delays showed no significant changes during the quiet and disturbed periods. The same finding was presented by [10] in their study on the African equatorial region (geomagnetic coordinates: 28.99°N–100.99°E). They observed two severe geomagnetic storm events in November 2004. Nonetheless, even a slight increase in forecasting results must be taken into consideration because it increases forecasting error when compared with the results obtained during the quiet periods and leads to a slight reduction in the accuracy of the forecasting model during disturbed periods.

5.10 ARIMA Model

The ARIMA model allows the production of a set of weighted coefficients that describe the ionosphere's behaviour or rate of change during the sample period. These coefficients can then be used to forecast future observations. The modelling of time-series observations focuses on three different phases: identification, estimation and application. Firstly, an appropriate ARIMA modelling strategy needs to be identified for a particular time-series. Then, the model parameters are estimated and refined. Finally, the model can be applied to predict future values. The mathematical equations of the ARIMA model can be written as

$$y_t = \Phi_1 y_{t-1} + \Phi_2 y_{t-2} + \ldots + \Phi_p y_{t-p} + u_t - \Theta_1 u_{t-1}$$
$$- \Theta_2 u_{t-2} - \Theta_2 u_{t-2} - \ldots - \Theta_q u_{t-q} \qquad (5.26)$$

$$\Phi_p(B) = \left(1 - \Phi_1 B - \ldots - \Phi_p B^p\right) \qquad (5.27)$$

$$\Theta_q(B) = \left(1 - \Theta_1 B - \ldots - \Theta_q B^q\right) \qquad (5.28)$$

$$\Phi_p(B)(1 - B)^d y_t = \Theta_q B(u_t) \qquad (5.29)$$

where

Y_t forecast variable.
B back shift operator.
ϕ AR coefficient.
Θ MA coefficient.
u_t white noise times series.

5.11   Comparison of the Holt–Winter and the ARIMA Models

The result obtained from the Holt–Winter was compared with that acquired with the ARIMA model during selected quiet days from 14 to 17 October 2011 and disturbed days from 23 to 28 October 2011 over the UKM station. The diurnal variations of the forecasted ionospheric delays obtained from the Holt–Winter and the ARIMA model during the quiet and disturbed periods are shown in Fig. 5.16. In this figure, the ionospheric delay is provided on the vertical axis, and the quiet and disturbed periods are given on the horizontal axis. The quiet period is shown in the top panel, and the disturbed period is depicted in the bottom panel. As illustrated in Fig. 5.16, the diurnal ionospheric delay forecasted for the quiet period by using the Holt–Winter model ranged from 0.7 m to 7.2 m and that forecasted by the ARIMA model ranged from 0.5 m to 8 m. The forecasting error during the disturbed period was higher than that during the quiet period. The forecasting errors during the disturbed period of the Holt–Winter and ARIMA models ranged between 0.6–8.5 and 0.7–9.7 m, respectively.

The MAPE variations for the Holt–Winter and ARIMA models during the quiet and disturbed periods were estimated to test the accuracy of the forecasting models and shown in Fig. 5.17. In this figure, the MAPE value is presented on the vertical axis, and the quiet and disturbed days are given on the horizontal axis. During the quiet period, the maximum MAPE value of the Holt–Winter was 5.2% and that of the ARIMA was 19.5%. The maximum MAPE values of the Holt–Winter and ARIMA models during disturbed periods were 6.6% and 26.7%, respectively. Therefore, the Holt–Winter provided a highly accurate forecast during the quiet and disturbed periods, whereas the ARIMA model can give a good forecast during the quiet period and a reasonable forecast during the disturbed period. As could be deduced from the analytical and forecasting results, the Holt–Winter model provided better estimates of ionospheric delay during the quiet and disturbed periods with lower forecasting error than the ARIMA model.

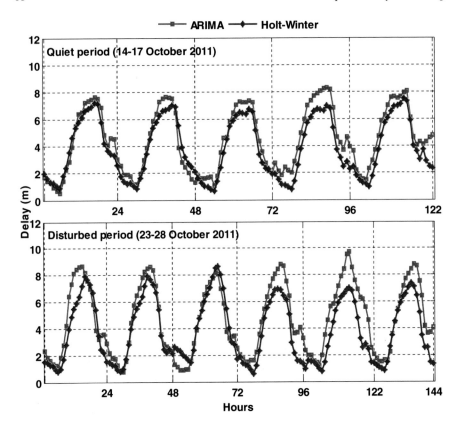

Fig. 5.16 Comparison of the diurnal ionospheric delays during the quiet and disturbed period forecasted by using the Holt–Winter and the ARIMA models

5.12 Summary

The Holt–Winter method was explained and modified to provide forecasting results with high accuracy. The MAPE was selected to test the accuracy of the method. The method provided good forecasts as long as the MAPE was less than 20% and highly accurate forecasts if the MAPE was equal or less than 10%. The ARIMA model, defaults, equations and details were discussed. A dual-frequency GISTM receiver was installed at the UKM and Langkawi stations. Data recording and processing by the modified Holt–Winter method were described in detail. As described in this chapter, the performances of the A-HW and M-HW models during 15 months at the UKM station were compared, and the accuracy of each model was tested. In general, the ionospheric delay trends forecasted by both models were similar to the actual delay with a comparatively small error of 0–0.2 m. The M-HW model gave a more accurate forecast than the A-HW model. Thus, the M-HW model was selected as the model for forecasting ionospheric delay over Malaysia. Furthermore, the ionospheric

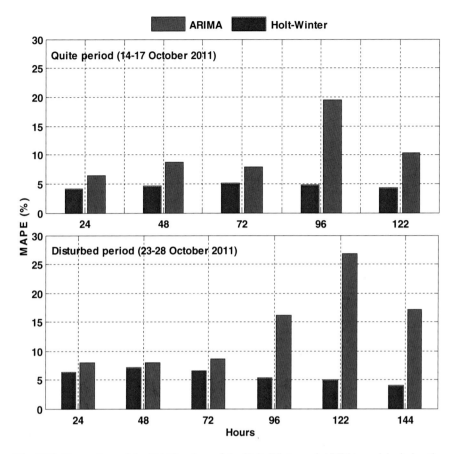

Fig. 5.17 Comparison of the MAPE values of the Holt–Winter and ARIMA models during the quiet and disturbed periods

delay over the UKM and Langkawi stations for the year 2011 was forecasted by using the M-HW model, and the result was validated with the actual ionospheric delay. The results showed that ionospheric delay had a minimum value during sunrise at 5:00 LT and a maximum value during the post noon period of 13:00–17:00 LT. The forecasted trends were similar but were slightly underestimated relative to the actual ionospheric delay observed over the UKM and Langkawi stations for the entire year. The highest monthly values for ionospheric delay were observed in March, and the lowest values were observed in July. Ionospheric delay was higher over the UKM station than over the Langkawi station. It was lowest during the summer season and highest during the equinox season. The maximum forecasted ionospheric delay for both stations was observed during the geomagnetic storm on 25 October. Comparing the MAPE values for the quiet and disturbed periods revealed that the MAPE for disturbed periods was slightly higher than that for the quiet periods. However, given that the MAPE was less than 10%, the forecasting results were considered to be highly accurate. The MAPE for the Langkawi station was higher than that for the UKM station. The accuracy of the Holt–Winter method for the quiet period was approximately 98% and that for the disturbed period was approximately 93%. In addition, the results of the Holt–Winter method for the quiet and disturbed periods were compared with those of the statistical ARIMA model. The Holt–Winter method had better forecasting results for the quiet and disturbed periods than the ARIMA model. Hence, the study proved the effectiveness of the Holt–Winter method in forecasting ionospheric delay. This model gave good results for diurnal, monthly and seasonal variations during quiet and disturbed periods. These results could be useful in representing model errors by using data assimilation. Consequently, this model could help improve ionospheric models and to correct ionospheric delay errors to improve the accuracy and the performance of GPS positioning in equatorial regions, particularly Malaysia.

Frequently, GPS frequency delays are unknown. Thus, ionospheric delay cannot be calculated properly. Under this condition, ionospheric VTEC modelling is used to present the ionospheric delay. Therefore, the variability of ionospheric VTEC and model comparisons will be discussed in the next chapter.

References

1. S. Arunpold, N.K. Tripathi, V. Rajesh Chowdhary, D.K. Raju, Comparison of GPS-TEC measurements with IRI-2007 and IRI-2012 modeled TEC at an equatorial latitude station, Bangkok, Thailand. J. Atmos. Solar Terr. Phys. **117**, 88–94 (2014). https://doi.org/10.1016/j.jastp.2014.06.001
2. S. Basu, S. Basu, F.J. Rich, K.M. Groves, E. MacKenzie, C. Coker, Y. Sahai, P.R. Fagundes, F. Becker-Guedes, Response of the equatorial ionosphere at dusk to penetration electric fields during intense magnetic storms. J. Geophys. Res.: Space Phys. **112**(8) (2007). https://doi.org/10.1029/2006JA012192
3. F.M. Dujanga, J. Mubiru, B.F. Twinamasiko, C. Basalirwa, T.J. Ssenyonga, Total electron content variations in equatorial anomaly region. Adv. Space Res. **50**(4), 441–449 (2012)
4. H.A. da Silva, P. de Oliveira Camargo, J.F. Galera Monico, M. Aquino, H.A. Marques, G. De Franceschi, A. Dodson, Stochastic modelling considering ionospheric scintillation effects on GNSS relative and point positioning. Adv. Space Res. **45**(9), 1113–1121 (2010). https://doi.org/10.1016/j.asr.2009.10.009
5. C.J. Davis, M.N. Wild, M. Lockwood, Y.K. Tulunay, Ionospheric and geomagnetic responses to changes in IMF B z : a superposed epoch study. Ann. Geophys. **15**(2), 217–230 (1997). https://doi.org/10.1007/s00585-997-0217-9
6. N.A. Elmunim, M. Abdullah, S.A. Bahari, Characterization of ionospheric delay and forecasting using GPS-TEC over equatorial region, Malaysia. Ann. Geophys. **63**(2), PA211 (2020). https://doi.org/10.4401/ag-8066
7. N.A. Elmunim, M. Abdullah, A.M. Hasbi, S.A. Bahari, Comparison of statistical holt-winter models for forecasting the ionospheric delay using GPS observations. Indian J. Radio Space Phys. **44**(1), 28–34 (2015). https://www.scopus.com/inward/record.uri?eid=2-s2.0-84930336944%7B&%7DpartnerID=40%7B&%7Dmd5=5d40c88d1af354d2bc2d5f0b11fc1d49
8. N.A. Elmunim, M. Abdullah, A.M. Hasbi, S.A. Bahari, *Comparison of GPS TEC Variations with Holt-Winter Method and IRI-2012 Over Langkawi* (Advances in Space Research, Malaysia, 2016). https://doi.org/10.1016/j.asr.2016.07.025
9. P. Galav, S. Sharma, S.S. Rao, B. Veenadhari, T. Nagatsuma, R. Pandey, Study of simultaneous presence of DD and PP electric fields during the geomagnetic storm of November 7–8, 2004

and resultant TEC variation over the Indian Region. Astrophys. Space Sci. **350**(2), 459–469 (2014). https://doi.org/10.1007/s10509-014-1792-3

10. J.B. Habarulema, L.A. McKinnell, D. Burešová, Y. Zhang, G. Seemala, C. Ngwira, J. Chum, B. Opperman, A comparative study of TEC response for the African equatorial and mid-latitudes during storm conditions. J. Atmos. Solar Terr. Phys. **102**, 105–114 (2013). https://doi.org/10.1016/j.jastp.2013.05.008

11. M.C. Kelley, D. Kotsikopoulos, T. Beach, D. Hysell, S. Musman, Simultaneous Global Positioning System and radar observations of equatorial spread F at Kwajalein. J. Geophys. Res. Space Physics **101**(A2), 2333–2341 (1996). https://doi.org/10.1029/95JA02025

12. A. Komjathy, R. Langley, An assessment of predicted and measured ionospheric total electron content using a regional GPS network. *Of the National Technical Meeting of ....* (1996). http://gauss.gge.unb.ca/grads/attila/papers/ionntm/ion96ntm.pdf

13. S. Kumar, K. Patel, A.K. Singh, TEC variation over an equatorial and anomaly crest region in India during 2012 and 2013. GPS Solutions **20**(4), 617–626 (2016). https://doi.org/10.1007/s10291-015-0470-4

14. C.D. Lewis, *Industrial and business forecasting methods: A practical guide to exponential smoothing and curve fitting.* (Butterworth-Heinemann, 1982)

15. S. Makridakis, R.J. Hyndman, S.C. Wheelwright, *Forecasting : Methods and Applications* (Wiley, 1998)

16. O.J. Olwendo, P. Baki, C. Mito, P. Doherty, Characterization of ionospheric GPS Total Electron Content (GPS-TEC) in low latitude zone over the Kenyan region during a very low solar activity phase. J. Atmos. Solar Terr. Phys. **84–85**, 52–61 (2012). https://doi.org/10.1016/j.jastp.2012.06.003

17. S. Panda, S. Gedam, G. Rajaram, Study of Ionospheric TEC from GPS observations and comparisons with IRI and SPIM model predictions in the low latitude anomaly Indian subcontinental region. Adv. Space Res. (2015) http://www.sciencedirect.com/science/article/pii/S0273117714005687

18. P.V.S. Rama Rao, S. Gopi Krishna, J. Vara Prasad, S.N.V.S. Prasad, D.S.V.V.D. Prasad, K. Niranjan, Geomagnetic storm effects on GPS based navigation. Ann. Geophys. **27**(5), 2101–2110 (2009). https://doi.org/10.5194/angeo-27-2101-2009

19. G.S. Valley, *GSV4004B GPS Ionospheric Scintillation & TEC Monitor (GISTM) User's Manual.* (Los Altos, 2007)

20. S. Zou, A. Ridley, M. Moldwin, Multi-instrument observations of SED during 24–25 October 2011 storm: implications for SED formation processes. Journal of (2013). http://onlinelibrary.wiley.com/doi/https://doi.org/10.1002/2013JA018860/full

Chapter 6
Modelling the Ionospheric VTEC and Forecasting

Ionospheric VTEC is the main ionosphere parameter with a substantial effect on radio wave propagation, thus causing delay errors in GPS signals. The ionospheric VTEC is normally used to present ionospheric delay regardless of the values of the GPS frequency delay when unknown. investigate the characteristics of ionospheric behaviour and errors in the equatorial region is important to Identify the most effective model in terms of accuracy levels to reduce GPS positioning errors. The use of the modified statistical Holt–Winter model to forecast ionospheric VTEC in different latitudes during geomagnetically quiet and disturbed periods is discussed in this chapter. Ionospheric delay due to geomagnetic storm disturbance, which is regarded as one of the most important phenomena that can significantly affect TEC and therefore ionospheric delay, was investigated and forecasted during different geomagnetic storm events in different years and at different latitudes. The Holt–Winter model was proven to be effective in forecasting ionospheric delay, providing accurate forecasting results for periods of disturbance. Given that the International Reference Ionosphere model is widely used and frequently updated, a recent version of the IRI-2016 model was compared with IRI-2012 to evaluate and investigate model improvements. The actual diurnal, monthly and seasonal variations of ionospheric VTEC were compared with the results of the Holt–Winter and IRI models by using three different topside electron density options, namely, IRI-2001, IRI01-corr and NeQuick, to validate the efficiency of the Holt–Winter method. Then, the performance and efficiency of the Holt–Winter method and the IRI-2012 and IRI-2016 topside options during quiet and disturbed periods at different latitudes were evaluated and tested.

6.1 GPS Receiver and Data Processing

TEC data can be collected from a different type of receivers. As discussed in this section and shown in Fig. 6.1, GPS-TEC data collected from the dual-frequency

© The Author(s), under exclusive license to Springer Nature Singapore Pte Ltd. 2021
N. A. Elmunim and M. Abdullah, *Ionospheric Delay Investigation and Forecasting*,
SpringerBriefs in Applied Sciences and Technology,
https://doi.org/10.1007/978-981-16-5045-1_6

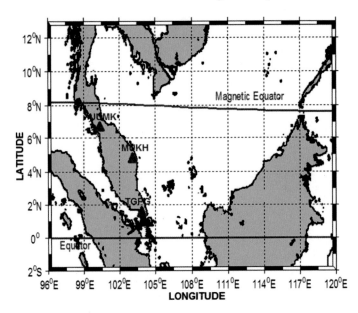

Fig. 6.1 Map depicting the geographical locations of the GPS receivers over the UUMK, MUHK and TGPG stations

GPS receivers over Universiti Utara Malaysia Kedah (UUMK) (geographic coordinates 4.62°N–103.21°E, geomagnetic coordinates: 5.64°N–174.98°E), Mukhtafibillah (MUKH) (geographic coordinates: 6.46°N–100.50°E, geomagnetic coordinates: 3.32°S–172.99°E) and Tanjung Pengerang (TGPG) (geographic coordinates: 1.36°N–104.10°E, geomagnetic coordinates: 8.43°S-176.53°E) were obtained from the Department of Survey and Mapping Malaysia (DSMM). These stations were chosen to determine the difference in ionospheric delays at different latitudes to understand the effect of the equator and the magnetic equator on the trend of the STEC. The STEC was determined from the receiver-independent exchange (RINEX) format observation files obtained from the DSMM. The GPS RINEX files were processed by using the GPS-TEC analysis software developed by Gopi, an institute for scientific research in Boston College [18]. This software is available for free on the Internet at http://seemala.blogspot.com. The GPS-TEC analysis software used the phase and code measurements for the L1 and L2 frequencies to calculate relative STEC values. The absolute values of the STEC were then obtained by including the differential satellite biases published by the University of Bern and the receiver bias that was calculated by minimising the TEC variability between 02:00 and 06:00 LT [10]. The TEC values were generated in ASCII files, which also included information on the time, elevation angle, azimuth angle, IPP-longitudes and IPP-latitudes. The VTEC value was computed by averaging the TEC for individual satellites in view. The elevation cut-off point of 20° was used to minimise the effect of the multipath phenomenon on the GPS data. The software program was used to process the RINEX observation files. It needed the observation date from the RINEX navigation file to estimate the azimuth and elevation angles of the satellites, which were subsequently needed for the VTEC calculation. The RINEX navigation file was downloaded automatically from the Internet into the original data observation folder, decompressed and subsequently used.

6.2 Ionospheric VTEC Variations and Forecasting

Data were collected from the dual-frequency GPS receivers of the DSMM over the UUMK, MUKH and TGPG stations to investigate the variations of ionospheric VTEC at different latitudes. The Holt–Winter model was used to forecast the VTEC data for each station individually. This approach could help understand the variability of the measured VTEC and the suitability of the Holt–Winter model to forecast the VTEC across different latitudes during geomagnetically quiet and disturbed periods. The data collected during March 2013 were used. On 18 March 2013, an intense geomagnetic storm affected the observed VTEC trend. The VTEC trend obtained from the observed GPS-TEC was compared with that forecasted by the Holt–Winter model, and the efficiency of the forecasting model during the quiet and disturbed periods was tested on the basis of MAPE error measurement.

6.2.1 Diurnal Variations

A typical quiet day with Kp ≤ 0 on 7 March 2013 was chosen to show the diurnal variations of the VTEC. In Fig. 6.2, the hourly median local time is plotted on the

Fig. 6.2 Diurnal hourly variation of the VTEC values measured over UUMK, MUKH and TGPG stations

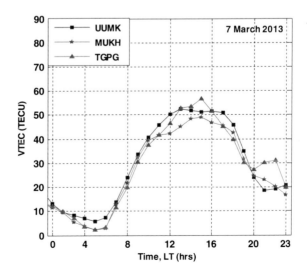

horizontal axis against the VTEC on the vertical axis. The overall trend of the VTEC for the UUMK, MUKH and TGPG stations indicated that the VTEC reached its minimum value during sunrise between 5:00–7:00 LT and then gradually increased to its maximum value during post noon hours between 13:00–17:00 LT. The behaviour of this trend was similar to that of the trends observed for the Langkawi and UKM stations as described earlier in Chap. 5. The diurnal variation of VTEC for most of the daily hours was higher at the UUMK station than at other stations. Comparing the VTEC values across all the investigated stations revealed that the VTEC values during 03:00–12:00 and 17:00–19:00 LT were higher at UUMK than at the MUKH and TGPG stations, whereas the VTEC values at 15:00, 21:00 and 22:00 LT were higher at the TGPG station than at other stations. The VTEC at peak hours from 12:00–16:00 LT was lower at the MUKH station than at other stations. The variability of VTEC at these stations was more prominent during peak post noon hours than at other hours. As shown in Fig. 6.1, the UUMK station is located near the magnetic equator. Hence, the slightly higher values of VTEC at the UUMK station than at other stations could be explained in reference to [7], who indicated that ionospheric VTEC increases with the increase in solar activity. The VTEC data were recorded during a period of high solar activity in 2013. This period was regarded as the maximum phase period of the solar cycle 24. During this period, the global vertical F-region drifts became large. In addition, the influence of the enhanced $E \times B$ vertical drift on the equator led to the rise in F-region plasma [20]. This effect delayed the decay time of the plasma, thus resulting in the abnormally slightly higher VTEC values recorded at the UUMK station than at other stations.

6.2.2 Daily Variations During the Quiet Period

The diurnal variations of the daily VTEC during the quiet period of 6–11 March 2013 at the UUMK, MUKH and TGPG stations were determined by using GPS-TEC values and the Holt–Winter model as shown in Fig. 6.3. In this figure, the VTEC (TECU) is shown on the vertical axis and the quiet days are shown on the horizontal axis. As illustrated in Fig. 6.3, the GPS-TEC measurement for MUKH was approximately 54 TECU and that for the UUMK and TGPG stations was 58 TECU. Comparing the trends of the Holt–Winter model with those of the GPS-TEC measurements revealed that the Holt–Winter predictions showed good agreement with the GPS-TEC measurements. However, a slight underestimation for peak hours on some days was observed. The maximum differences between the GPS-TEC measurements and the Holt–Winter results for MUKH, UUMK and TGPG stations were 3.8, 4.1 and 4.3 TECU, respectively.

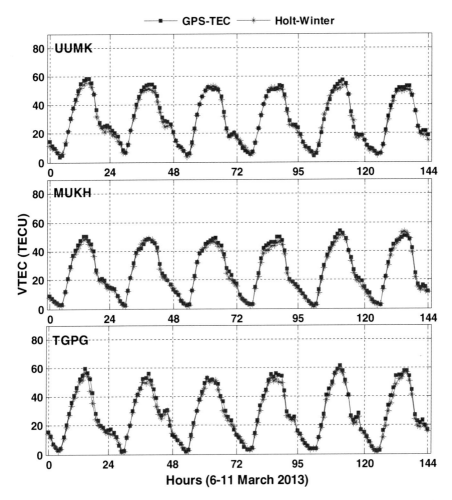

Fig. 6.3 Daily variations of the VTEC from GPS-TEC measurements and the Holt–Winter model forecasts for quiet periods

6.2.3 Daily Variations During the Disturbed Period

The period of 15–20 March 2013 was selected to investigate the variations in the VTEC over the UUMK, MUKH and TGPG stations during geomagnetic storm disturbance. The selected period encompassed 3 days before the storm and 2 days after the storm. An intense geomagnetic storm on 18 March 2013 caused a sharp reduction in Dst values, which reached their minimum value of −132 nT. This disturbance increased the VTEC trend, and its effect during the time interval of 9:00–15:00 LT was estimated. This geomagnetic storm event will be discussed in further detail in Sect. 6.8. Figure 6.4 shows the variations of the GPS-TEC measurements and the

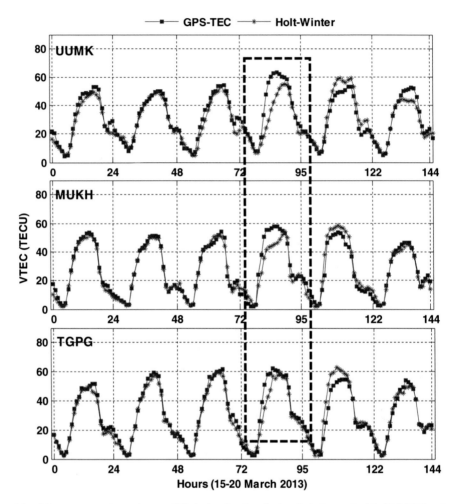

Fig. 6.4 Daily variations of the VTEC during the disturbed period observed with GPS-TEC and predicted with the Holt–Winter model

Holt–Winter model forecasts during the disturbed period. The GPS-TEC measurements steadily increased 3 days before the storm until reaching their maximum value on the day of the storm on 18 March and then decreased on the days after the storm. The highest GPS-TEC values for the UUMK, MUKH and TGPG stations were approximately 58, 63 and 62 TECU, respectively. The measured VTEC (GPS-TEC) and values predicted by using the Holt–Winter model 3 days before the geomagnetic storm were in good agreement. However, the VTEC value forecasted by the Holt–Winter model for 18 March was underestimated relative to the GPS-TEC measurement. The Holt–Winter model also overestimated the VTEC for all stations for the day after the geomagnetic storm (19 March).

6.2.4 Evaluating the Performances of the Holt–Winter Model During the Quiet and Disturbed Periods

MAPE error measures were used to evaluate the accuracy of the Holt–Winter model. The MAPE values for the UUMK, MUKH and TGPG stations during the quiet and disturbed periods are shown in Fig. 6.5, where the MAPE (%) values are presented on the vertical axis, the days of the quiet period are shown in the top panel and the days of the disturbed period are shown in the bottom panel. During the quiet period, the MAPE for MUKH, UUMK and TGPG fell in the ranges of 3.5–4.4%, 3.0–3.7% and 3.2–4.2%, respectively. During the disturbed period, the MAPE for MUKJ, UUMK and TGPG fell in the ranges of 3.5–9.4%, 3.8–8.6% and 3.6–7.6%, respectively. For all of the stations, the maximum MAPE during the quiet period was 4.4%, whereas the maximum MAPE during the disturbed period was 9.4%.

Therefore, the Holt–Winter is a regional method that is suitable and reliable for estimating and forecasting ionospheric VTEC at different latitudes over Malaysia given that it produced an accurate forecasting result for quiet and disturbed periods.

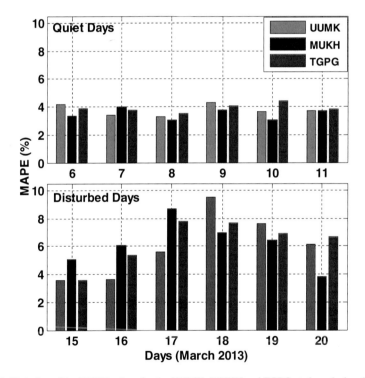

Fig. 6.5 Variation of the MAPE values for the UUMK, MUKH and TGPG stations during the quiet and disturbed periods

6.3 IRI-2012 Model

The IRI model is one of the most widely utilised effective empirical models for forecasting ionospheric VTEC variability worldwide. The IRI model has been upgraded, and the update site is accessible from the IRI website at http://IRI.gsfc.nasa.gov.

The IRI-2012 model incorporates considerable enhancement for the representation of electron density and some other parameters (see Sect. 4.1). The TEC ranges from altitudes of 60 km to 2000 km. The IRI-2012 model integrates electron density from the lower to the upper specified boundaries and provides the TEC [5]. The options for bottom-side electron density include ABT-2009, Gul-1987 and Bil-2000, and those for top-side electron density are IRI01-corr, IRI-2001 and NeQuick. The various top-side electron density options are expressed by various mathematical functions. The IRI-2012 storm model was developed to describe the behaviour of an average storm and could be turned ON and OFF. In this research, the ABT-2009 option was selected by using the URSI and the F peak models to determine the VTEC with the IRI-2012 model. The NeQuick, IRI01-corr and IRI-2001 top-side electron density options were given in the model to generate the hourly values of the VTEC for a particular day of the year. The altitude upper boundary of the IRI-2012 model reached 1500 km, and the ON and OFF options were used in the storm model. The VTEC values estimated from the IRI-2012 model and GPS signals may vary because of plasmaspheric contribution. However, the plasmasphere is often ignored in the estimation and analysis of GPS-TEC data and in comparison with the IRI-2012 model [11]

The percentage deviation (%Dev) was acquired from the variation of the VTEC for each month and season through approximation with Eq. (6.1) to validate the IRI model. The percentage deviation approach provides a good model for describing prediction model errors. Accuracy was calculated on the basis of the mean error (ME), which was obtained by using Eq. (6.2):

$$\% \text{ Dev} = \left(\frac{\text{VTEC}_y - \text{VTEC}_x}{\text{VTEC}_x} \right) \times 100 \qquad (6.1)$$

$$\text{ME} = \frac{1}{N} \sum_{t-1}^{N} (\text{VTEC}_{\text{model}} - \text{VTEC}_{\text{observed}}) \qquad (6.2)$$

where

$VTEC_x$ measured VTEC.
$VTEC_y$ modelled VTEC.
N number of total observations.
t time period.

6.4 Comparison of the Holt–Winter Model with the IRI-2012 Model

The VTEC derived from the GISTM receiver over the Langkawi station in the northern region of Malaysia (geographic coordinates: 6.19°N, 99.51°E; geomagnetic coordinates: 3.39°S–172.42°E) during 2014 was measured by using GPS-TEC and compared with the modelled VTEC by using the Holt–Winter and the IRI-2012 model with three top-side electron density options, namely IRI-2001, IRI01-corr and NeQuick, to verify and highlight the capability of the Holt–Winter model. Hourly, monthly and seasonal variations were compared to validate, compare and then determine the most effective model in terms of accuracy levels.

6.4.1 Diurnal Variations

The diurnal variations of the VTEC for the year 2014 obtained from GPS-TEC were compared with those acquired with the Holt–Winter and the IRI-2012 top-side options (IRI-2001, IRI01-corr and NeQuick). Figure 6.6 shows a typically quiet day with Kp <2. In this figure, the VTEC is presented on the vertical axis in TECU units, and the time is given on the horizontal axis. Generally, the overall VTEC trends obtained via GPS-TEC measurements and the Holt–Winter, IRI-2001, IRI01-corr and NeQuick models behaved similarly and reached their minimum value at sunrise between 5:00 and 6:00 LT. Then, they gradually increased until they reached their maximum value during 13:00–17:00 LT in the afternoon. They then decreased again at sunset. Chakraborty et al. [8] compared the VTEC for four equatorial to mid-latitude stations in 2012 measured by using GPS-TEC with that predicted by using IRI-2012 top-side options. The minimum diurnal variation was observed at predawn, whereas the maximum was observed after local noon. The variation decreased again

Fig. 6.6 Diurnal hourly variations of the measured and modelled VTEC

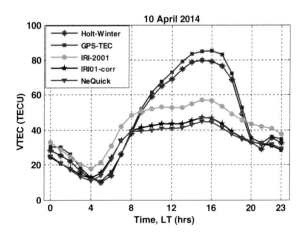

at night. As illustrated in Fig. 6.6, NeQuick and IRI01-corr showed similar trends during the day. All models underestimated the observed GPS-TEC value for the early morning hours of 2:00–5:00 LT. Then, the IRI-2001, IRI01-corr and NeQuick models overestimated the VTEC for the hours of 6:00–10:00 LT, whereas the result of the Holt–Winter model was in good agreement with the GPS-TEC measurement. The VTEC trends for peak hours (10:00–21:00 LT) forecasted by all models were underestimated relative to GPS-TEC measurements, and those for 22:00 LT modelled by the IRI-2001 were overestimated [9].

6.4.2   Monthly Variations

The hourly medians of the VTEC were calculated to obtain the monthly variability. As depicted in Fig. 6.7, the hourly medians were then plotted against time for each month to represent the monthly variations of the VTEC estimated from GPS-TEC measurements and that modelled using the Holt–Winter model and IRI-2012 with three top-side options. The VTEC for January predicted by IRI-2001, NeQuick, IRI01-corr and the Holt–Winter model was in good agreement with the GPS-TEC measurements. The VTEC for hours after noontime gradually increased from February to April. It then increased to its peak in March. All IRI-2012 top-side options significantly underestimated the VTEC for peak hours. The GPS-TEC measurement was 84 TECU, and the Holt–Winter, IRI-2001, IRI01-corr and NeQuick predictions were 80, 53, 42 and 41 TECU, respectively. The VTEC trend gradually decreased from May to August, attaining a minimum value in June. The GPS-TEC measurement was 41 TECU, and the Holt–Winter, IRI01-corr and NeQuick predictions were 40, 39 and 36 TECU, respectively. IRI-2001 provided the largest overestimation of approximately 50 TECU. The VTEC increased gradually from September until December. Comparison with GPS-TEC measurements revealed that IRI-2001, NeQuick and IRI01-corr underestimated the values for all months, whereas IRI-2001 overestimated the values for June, July and August. The Holt–Winter model had slight underestimations for all months of the year. Similar studies using the IRI model [23] compared the prediction performances of GPS-TEC and the IRI-2007 model for two stations in the equatorial region over India. For both stations, the GPS-TEC value for the daytime was underestimated and the GPS-TEC value for the night-time was overestimated. Kumar et al. [12] compared the performance of the GPS-TEC with that of the IRI-2007 model over three stations, namely Varanasi (geomagnetic coordinates: 16.10°N–156.79°E), Hyderabad (geomagnetic coordinates: 8.56°N–151.87°E) and Bangalore (geomagnetic coordinates: 4.27°N–150.63°E), India from 2007 to 2009. Their results indicated that the results of the IRI01-corr and the NeQuick models exhibited good agreement with those of the GPS-TEC for daytime hours in Varanasi and Hyderabad. The IRI-2001 model result was also in agreement with the GPS-TEC measurement for daytime hours in Bangalore. Kumar et al. [13] also stated that the results of the GPS-TEC and the IRI-2012 model for Singapore (geomagnetic coordinates: 8.62°S–175.99°E) in the equatorial region during 2010 were in

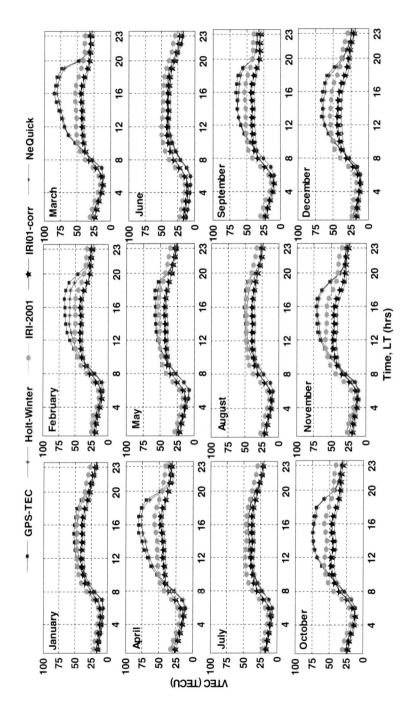

Fig. 6.7 Monthly variation of the measured and modelled VTEC during 2014

close agreement, indicating that the IRI-2012 model offered better results than the IRI-2007 model. The results of this research showed that in the diurnal monthly variation of the GPS-TEC and the IRI-2012 model, the IRI01-corr and NeQuick model results demonstrated good agreement with GPS-TEC measurements, whereas the IRI-2001 model did not show any agreement with the GPS observations. In general, the validation of the Holt–Winter and IRI-2001, IRI01-corr and NeQuick results with GPS-TEC measurements showed that the Holt–Winter model provided better results for all months of the year than the IRI-2001, IRI01-corr and NeQuick models.

The monthly variability of the VTEC can be described by using the ratio of the %Dev of the monthly values. In Fig. 6.8, the %Dev of the VTEC for the year 2014 is plotted monthly and is presented on the vertical axis, and the hourly time is presented on the horizontal axis. Figure 6.8 shows that the %Dev increased significantly in the morning from 05:00 to 09:00 LT and reached its maximum peak at 06:00 LT. However, low %Dev values could be observed from 10:00 to 19:00 LT. A slight increase was then observed after 19:00 LT until 22:00 LT. For 05:00–09:00 LT, the %Dev provided by IRI-2001 was the highest, followed by those given by the IRI01-corr, NeQuick and then the Holt–Winter models. The sudden increase in the %Dev seen in all of the IRI-2012 top-side options (IRI-2001, IRI01-corr and NeQuick) could be attributed to the incapability of the IRI-2012 model to give a good result for the hours of 4:00–8:00 LT. Moreover, the model is still undergoing improvement. A close inspection in Fig. 6.9 illustrateed that the Holt–Winter and the IRI-2012 models exhibited lower %Dev for the period of 10:00–19:00 LT throughout 24 h. For June, the %Dev of the IRI-2001 model exhibited a significant increase compared with that of the night-time IRI01-corr, NeQuick and Holt–Winter models. The %Dev of the IRI01-corr and the Holt–Winter for 13:00 to 20:00 LT were close to zero as shown in Fig. 0. However, the monthly %Dev for 13:00–20:00 LT indicated that the IRI-2001 model had a lower %Dev than the IRI01-corr and NeQuick models during May, September, October and November. The difference in %Dev was positive when the modelled VTEC was overestimated as in the IRI-2001 model but was negative when the modelled VTEC was underestimated relative to the night-time GPS-TEC measurement as shown by the NeQuick, IRI01-corr and the Holt–Winter models. Adewale et al. [1] performed a similar study in 2009 by using the IRI-2007 model over Lagos, Nigeria (geomagnetic coordinates: 8.86°N–6.90°E) and observed that the maximum monthly %Dev occurred in March at 05:00 LT. In this work, however, the highest value of %Dev occurred in May as shown in Fig. 6.9. Comparing the %Dev values of the IRI-2001, NeQuick, IRI01-corr and the Holt–Winter models revealed that the IRI-2001 model had the largest %Dev for all months. The %Dev values increased significantly in May, June and July.

6.4.3   Seasonal Variations

The hourly median values were calculated for all days of the seasons to investigate the seasonal variations of the VTEC. The top panels in Fig. 6.10 illustrate that in

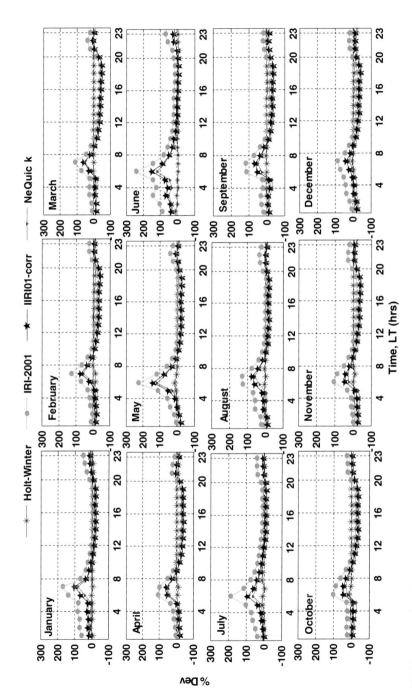

Fig. 6.8. Percentage deviation of the Holt–Winter method and the IRI-2012 top-side optionsduring the monthly variation for the year 2014

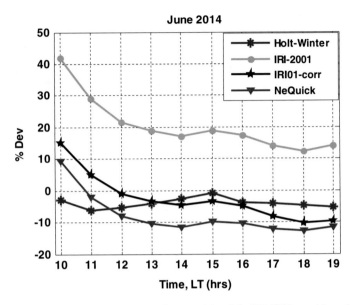

Fig. 6.9 Percentage deviation of the Holt–Winter model and the IRI-2012 top-side options during peak hours

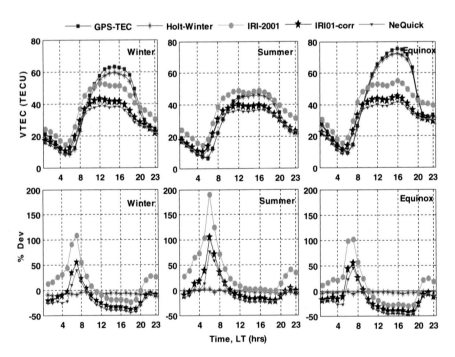

Fig. 6.10 Comparison of the measured and modelled seasonal VTEC values and the %Dev of seasonal variations during 2014

all the seasons observed, the maximum variation occurred from 9:00 LT to 19:00 LT. Seasonal VTEC values peaked during the equinox, showed the next-highest values during the winter and were lowest during the summer. The maximum VTEC measured by GPS-TEC during the equinox was 76 TECU. The next-highest VTEC values were obtained by the Holt–Winter model (72 TECU), IRI-2001 (55 TECU), IRI01-corr (44 TECU) and NeQuick (41 TECU). The VTEC values acquired with the IRI-2001 model were higher than those measured with GPS-TEC and were 49 and 47 TECU for the IRI-2001 model and GPS-TEC, respectively. The Holt–Winter model (46 TECU), IRI01-corr (39 TECU) and NeQuick (37 TECU) provided the next-highest values. The IRI-2001 model overestimated the VTEC values for all hours of summer season and overestimated the VTEC values for 4:00–10:00 LT during the winter and equinox. It underestimated the VTEC values for 11:00–20:00 LT relative to the GPS-TEC measurements. This behaviour was the same as the behaviours of NeQuick and IRI01-corr in the summer, winter and equinox. The Holt–Winter model underestimated the VTEC values for 11:00–20:00 LT and but was in good agreement with GPS-TEC for the remaining hours. As shown in the bottom side panels in Fig. 0, the highest seasonal %Dev was observed in summer mornings, whereas the %Dev trends during the equinox were relatively similar to those during the winter. During summer, the IRI-2001, IRI01-corr, NeQuick and Holt–Winter model showed lower %Dev for the peak hours of 11:00–19:00 LT, and the Holt–Winter and IRI-2001 models provided values close to zero. The %Dev was higher (with a negative value) during the equinox than during the summer and winter. The NeQuick, followed by IRI01-corr, had the largest %Dev. The IRI-2001 and then the Holt–Winter model had the next-highest %Dev. However, during 4:00–8:00 LT in the morning, the IRI-2001 had the highest %Dev value followed by NeQuick, IRI01-corr and then the Holt–Winter model.

Many related studies have been carried out. For example, Kumar et al. [14] compared the seasonal variation shown by the IRI-2012 model's three top-side options with those shown by GPS-TEC in the equatorial region over India between 2012 and 2013. NeQuick and IRI01-corr showed good agreement at all times in the summer but showed good agreement only between 5:00 and 13:00 UT in the equinox and the winter. During the equinox, NeQuick and IRI01-corr significantly underestimated the measured GPS-TEC values for 6:00–10:00 UT, whereas the IRI-2001 top-side option significantly overestimated the GPS-TEC measurement for all the seasons in 2012–2013. [2, 6, 13–15, 24] founded that the VTEC values modelled by the NeQuick and IRI01-corr were similar and generally tended to show good agreement with GPS-TEC measurements at all times, especially during the summer.

The results obtained through the diurnal, monthly and seasonal variations of the statistica Holt–Winter model and the empirical IRI-2012 model with top-side options were compared with GPS-TEC measurements. The Holt–Winter model estimated the VTEC value on the basis of the statistical mathematics applied to the previous time-series VTEC data and provided a better estimation of the VTEC than the empirical model, which was based on the empirical data derived from electron density measurements. The accuracies of the Holt–Winter model, IRI01-corr, NeQuick and IRI-2001 were 95%, 75%, 73% and 66%, respectively. As deduced from the comparative analysis of the Holt–Winter model, IRI-2001, NeQuick and IRI01-corr, the Holt–Winter model offered a better estimation of the VTEC than the IRI01-corr and NeQuick, whereas IRI-2001 showed poor predictive performance for the equatorial region.

6.5   IRI-2016 Model

This model is also continuously being updated by the scientific community [4]. The IRI-2016 model, the latest version released in 2017, incorporates considerable enhancement for electron density representation, as well as other parameters [3]. It is publicly accessible from the IRI website (http://IRI.gsfc.nasa.gov). As described in the following section, the performances of the IRI-2012 and IRI-2016 models during quiet and disturbed periods were compared by using a storm model. The ABT-2009 option for the bottom-side thickness parameter and the URSI option for the F2 peak density model was used to determine the VTEC from the IRI model using the IRI-2001, IRI01-corr and NeQuick top-side electron density options to generate the hourly values of the VTEC.

6.6   Performance of IRI-2016 and IRI-2012

The quiet day during a period of high solar activity in the year 2013 was selected and was regarded as the maximum of 24 h solar cycle to evaluate the performance of the IRI model. During this period, the global vertical F-region drifts were large and were coupled with enhanced $E \times B$ vertical drift on the equator [20, 21]. The hourly diurnal variation during the quiet day (7 March 2013) over the UUMK, MUKH and TGPG stations are shown in Fig. 6.11, with the first row representing the hourly variation over UUMK, MUKH and TGPG stations. The VTEC values (TECU) are displayed on the vertical axis, and the hours are represented by the horizontal axis. The IRI-2016 top-side options are denoted with a solid line; green, black and pink dotted lines represent IRI-2001, IRI01-corr and NeQuick, respectively, and green, black and pink lines indicate IRI-2001, IRI01-corr and NeQuick with IRI-2012 top-side options, respectively. The second row in Fig. 6.11 represents the difference between the corresponding values of GPS-TEC measurements and IRI models (DVTEC). Figure 6.11 illustrates that the IRI01-corr and NeQuick top-side options for IRI-2016 and IRI-2012 predictions for all times of the day, except during the peak hours from 11:00 to 18:00 LT, for all stations were overestimated relative to GPS-TEC measurements. The IRI-2001 option overestimated the GPS-TEC measurements for almost all the periods at TGPG stations and for most of the time at UUMK and MUKH stations. From 00:00–06:00 LT, the VTEC values observed from all top-side options (IRI-2001, IRI01-corr and NeQuick) were almost similar to the values of the IRI-2016 and IRI-2012 models. From 06:00 to 17:00 LT, the NeQuick option had a better

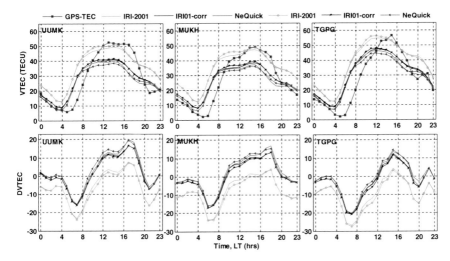

Fig. 6.11 Diurnal hourly variation and DVTEC of IRI-2016 and IRI-2012 top-side options over the UUMK, MUKH and TGPG stations. IRI-2016 values (IRI-2001, IRI01-corr and NeQuick) are plotted in solid lines, whereas IRI-2012 values (IRI-2001, IRI01-corr and NeQuick) are plotted in dotted lines

agreement with GPS-TEC than IRI-2012 (IRI-2001, NeQuick) and IRI-2016 (IRI-2001), whereas IRI-2012 (IRI01-corr) had comparatively better agreements with GPS-TEC than IRI-2016 (IRI01-corr). The IRI-2001 and IRI01-corr for IRI-2012 and IRI-2016 models displayed a similar trend from 18:00 to 23:00 LT, and IRI-2016 (NeQuick) showed better agreement with GPS-TEC than IRI-2012 (NeQuick). Patel et al. [17] investigated the performance of the IRI-2016 model for three stations (Bangalore, Hyderabad, Surat) in India and compared it with that of IRI-2012 by using IRI01-corr and NeQuick top-side options. The values of the IRI-2016 (IRI01-corr and NeQuick) for all stations, except for the morning hours, were similar to those of the IRI-2012 model, and the effect of the updated version (IRI-2016) was apparent only for the hours of 05:00–10:00 LT. Tariku (2018) discussed the improved performance of the IRI-2016 model for different regions of Ethiopia in the equatorial region. IRI-2012 generally performed better than IRI-2016 in capturing the diurnal GPS-TEC values. In this study, the highest DVTEC values for both IRI models were observed in the morning (06:00–07:00 LT) with the following TECU values: 23, 23 and 26 (IRI-2001),17, 15 and 20 (IRI01-corr); and 15, 15 and 20 (NeQuick) observed in IRI-2016 for each of the UUMK, MUKH and TGPG stations, respectively. IRI-2012 gave TECU values of 23, 23 and 27 (IRI-2001); 17, 16 and 20 (IRI01-corr) and 15, 19 and 20 (NeQuick) for the UUMK, MUKH and TGPG stations, respectively. In general, the DVTEC values for IRI-2016 and IRI-2012 models derived from the IRI0-corr model were lower than those derived from NeQuick and IRI-2001.

6.7 Ionospheric VTEC Modelled During Quiet Periods

The daily diurnal variations of VTEC during quiet periods with ascending and high solar activities in 2011, 2012 and 2013 were selected to investigate the performance of the ionospheric models for the UUMK station. Quiet days during 14–17 October 2011, 17–22 March 2012 and 7–11 March 2013 were selected. October and March are regarded as equinox months that have the highest range of VTEC values. During these months, the equinox meridional wind blows from the equator to the polar regions, subsequently causing a high ionisation peak value. The diurnal variations of the measured and modelled VTEC values obtained by using IRI-2016 (IRI-2001, IRI01-corr and NeQuick), IRI-2012 (IRI-2001, IRI01-corr and NeQuick) and the statistical Holt–Winter model over 5 quiet days are shown in Fig. 6.12. The GPS-TEC attained its maximum peak value during the post noon period and its minimum value during sunrise in each of the ascending and high solar activity periods in 2011, 2012 and 2013. The overall measured and modelled VTEC values were similar to the minimum values at sunrise and gradually increased until peaking at post noon and then decreasing at night. For the IRI-2016 and IRI-2012 (IRI-2001, IRI01-corr and NeQuick) models, the minimum values were attained at 04:00 LT and the maximum values were attained between 08:00–16:00 LT. The Holt–Winter model attained its minimum values from 04:00–06:00 LT and maximum values between 13:00–17:00 LT. Tariku (2018) compared the VTEC values obtained for Ethiopia in 2013 by

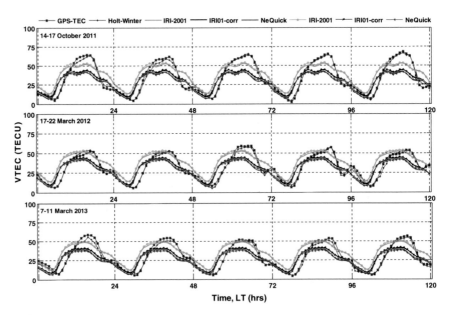

Fig. 6.12 Daily variation of the measured and modelled VTEC for quiet periods. The IRI-2016 (IRI-2001, IRI01-corr and NeQuick) model values are plotted by using solid lines and the IRI-2012 (IRI-2001, IRI01-corr and NeQuick) model values are plotted by using dotted lines

using the IRI-2016 top-side options. The minimum value was observed at predawn, the maximum value was observed at post noon and the VTEC declined again at night. The VTEC values estimated by IRI-2016, IRI-2012 (IRI-2001, IR01-corr and NeQuick) and the Holt–Winter models for peak hours between 11:00 to 17:00 LT were underestimated relative to those provided by GPS-TEC. All the models showed this underestimation during the period of low solar activity in 2011. For the remaining periods, the IRI-2001 prediction overestimated GPS-TEC measurements, whereas the predictions of IRI01-corr, NeQuick and the Holt–Winter models were consistent with GPS-TEC measurements for all the periods for 2011, 2012 and 2013 as compared with those of the IRI-2001 top-side option.

The IRI-2016 (IRI-2001, IRI01-corr and NeQuick) model had better prediction values for 2011 than IRI-2012. Similarly, the IRI-2016 (IRI-2001, NeQuick) model displayed better prediction values for 2012 than the other models. IRI-2012 (IRI-2001) showed a marginally better prediction for 21–22 March than IRI-2016. IRI-2016 (NeQuick) showed a better prediction for 2013 than the IRI-2012 model. However, from 20:00–23:00 LT during 2011, 2012 and 2013, the IRI-2012 (IRI-2001, IRI01-corr) model displayed better agreements with the GPS-TEC than IRI-2016. Tariq et al. [22] compared the performance of IRI-2012 and IRI-2016 for Pakistan and showed similar results, whereby IRI-2016 performed better than IRI-2012. However, Patel et al. [17] compared the performance of IRI-2016 and IRI-2012 for India and showed that the VTEC modelled by both IRI models were similar. Additionally, [19] investigated the improvements of the IRI model for Ethiopia and reported that IRI-2012 generally performed better than IRI-2016. Consequently, the performance of the IRI-2016 varied with different latitudes and regions.

The Holt–Winter model generally showed a good agreement with GPS-TEC during the investigated periods. Nevertheless, the modelled VTEC trends for 2012 and 2013 were slightly better than those for 2011. The IRI-2012 NeQuick model displayed the maximum underestimation of GPS-TEC measurements for the periods of 2011, 2012 and 2013. For the IRI-2016 model, NeQuick and IRI01-corr displayed a similar trend with the maximum underestimated values as compared with the IRI-2001 model, which showed maximum overestimation for IRI-2012 and IRI-2016 models for all the years investigated.

The differences between the measured and modelled VTEC values for selected quiet days in 2011, 2012 and 2013 are illustrated in Fig. 6.13. The hourly median local time was plotted on the horizontal axis against the DVTEC, which is represented by the vertical axis. The overestimation of the modelled VTEC was indicated by the positive difference in DVTEC values, whereas negative DVTEC values were acquired when the modelled VTEC was underestimated relative to the GPS-TEC measurement. The maximum DVTEC obtained from IRI-2016, IRI-2012, and Holt–Winter models presented in Table 6.1.

All models yielded the highest DVTEC values for the period of low solar activity in 2011. The Holt–Winter model showed a low DVTEC, and IRI01-corr had a better agreement with GPS-TEC and lower DVTEC than IRI-2001 and NeQuick for the IRI-2016 and IRI-2012 models.

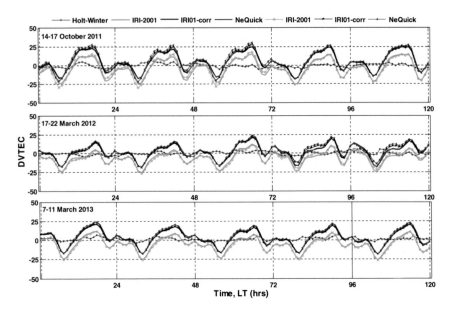

Fig. 6.13 Daily variation of DVTEC for the quiet period. The IRI-2016 (IRI-2001, IRI01-corr and NeQuick) model values are plotted with solid lines and the IRI-2012 (IRI-2001, IRI01-corr and NeQuick) model values are plotted with dotted lines

Table 6.1 The maximum DVTEC of IRI-2016 and IRI-2012 top-side options, and Holt-Winter model during quiet periods

Period	2011		2012		2013	
IRI model	IRI-2016	IRI-2012	IRI-2016	IRI-2012	IRI-2016	IRI-2012
IRI-2001	28	30	21	23	25	26
IRI01-corr	37	39	27	29	22	23
NeQuick	41	42	21	23	25	26
Holt-Winter	7		6		7	

6.8 Ionospheric VTEC Modelling During Disturbed Periods

The VTEC for three different geomagnetic storm events in October 2011, March 2012 and March 2013 was estimated to investigate and compare the IRI-2016 and IRI-2012 top-side options and to evaluate the performance of the Holt–Winter model during geomagnetic storm disturbance. The Dst value sharply decreased and attained its minimum value of approximately − 134, − 131 and − 132 nT during the main phases of the intense geomagnetic storms on 25 October 2011, 9 March 2012 and 18 March 2013, respectively. Figure 6.14 depicts the daily variation of the Dst values during October 2011, March 2012 and March 2013. It also indicates how the Dst

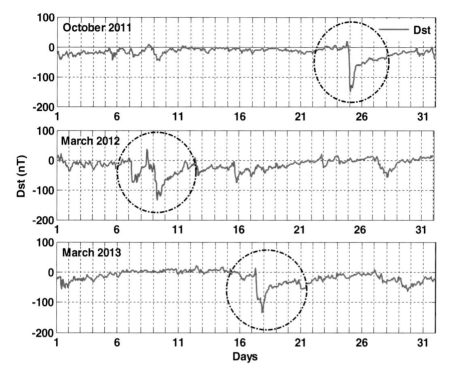

Fig. 6.14 Variations of the Dst index for October 2011, March 2012 and March 2013

values were affected by storm disturbances in three different years. In this figure, the Dst values are displayed on the vertical axis, and the days of the month are represented by the horizontal axis. A sudden storm commencement was observed on 25 October 2011 at 8:00 LT (24 October 2011, 24:00 UT) followed by the initial phase of the storm. The main phase of the storm occurred at 10:00 LT (2:00 UT) on 25 October, during which the Dst dropped rapidly to −134 nT. A two-step gradual storm commencement was observed in March 2012. Specifically, the storm started on 7 March and later entered the main phase on 8 March. A minimum Dst value of approximately −95 nT was observed. The short-lived recovery phase then followed. The second main phase occurred on 9 March with a minimum Dst of −131 nT. The recovery phase commenced on 10 March. The storm that occurred in March 2013 appeared to consist of two phases wherein the Dst decreased to −89 nT at 18:00 LT (10:30 UT) during the main phase and remained at the same level for a few hours before dropping rapidly again down to −132 nT on 18 March at 4:30 LT (20:30 UT, 17 March). The recovery phase started thereafter on 20 March.

The sudden decrease in Dst increased GPS-TEC measurements. VTEC values were measured for 7 days, including the day of the storm and 3 days before and after the storm. The maximum value of GPS-TEC was observed during the day of the geomagnetic storm and was approximately 80 TECU on 25 October 2011, 67 TECU on 9 March 2012 and 63 TECU on 18 March 2013. As illustrated in Fig. 6.15, the effects of the geomagnetic storm on the measured GPS-TEC data commenced 1 day before the storm (initial phase), whereby the data showed a slight increase and then steadily increased until the next day after the storm (recovery phase). The GPS-TEC data collected during this disturbed period were compared with the forecasting results of the IRI-2016 and IRI-2016 (IRI-2001, IRI01-corr and NeQuick) top-side options and the Holt–Winter model. However, to model the VTEC data for the geomagnetic storm event by using IRI-2016 and IRI-2012 (IRI-2001, IRI01-corr and NeQuick) top-side options, the VTEC data had to be estimated when the IRI storm model was switched ON and OFF. The modelled VTEC data from IRI-2016 and IRI-2012 models were not influenced by the ON or OFF switching of the storm model under geomagnetic storm conditions. Consequently, in all of the investigated periods, the IRI-2012 (IRI-2001, IRI01-corr and NeQuick) top-side options did not show any response during the disturbed period and were unaffected by the enhancement in VTEC during the geomagnetic storm.

The IRI-2016 (IRI-2001, IRI01-corr and NeQuick) top-side options had an insignificant response to the geomagnetic storm events in 2011, 2012 and 2013. In contrast to that during a geomagnetically quiet day, the IRI-2001 model showed a slight response to the geomagnetic storm at 02:00–20:00 LT in October 2011. By contrast, the IRI01-corr and NeQuick models showed a response at 02:00–05:00 LT and 11:00–17:00 LT. During the geomagnetic storm in March 2012, all top-side options displayed a slight enhancement at 08:00–17:00 LT, whereas the geomagnetic

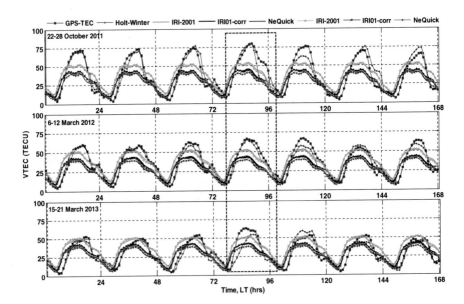

Fig. 6.15 Daily variation of the measured and modelled VTEC values during the disturbed period. The IRI-2016 (IRI-2001, IRI01-corr and NeQuick) model values are plotted in solid lines and the IRI-2012 (IRI-2001, IRI01-corr and NeQuick) model values are plotted in dotted lines

storm in March 2013 resulted in slight enhancements in the results of all the top-side options for the whole day. Since the Holt–Winter model used the data from the previous date to predict the following day, the increments in the GPS-TEC measurements during the storm period affected the performance of the Holt–Winter model, wherein modelling and forecasting errors increased during the periods of disturbance. The predictions of the Holt–Winter model varied between a slight overestimation and underestimation relative to the GPS-TEC data during the disturbed period and showed a clear underestimation relative to the GPS-TEC measurements for the storm days on 25 October 2011, 9 March 2012 and 18 March 2013. This underestimation occurred due to the sudden increase in GPS-TEC measurements during the geomagnetic storm. However, in different equatorial regions, some researchers had shown that IRI-2016 did not respond to the effects of geomagnetic storms, whereas others found a weak response. Mengistu et al. [16] reported that the IRI-2016 model had a slight response during a geomagnetic storm over the African region in March 2013. Tariku [20] investigated the effects of a geomagnetic storm on the measured VTEC values for March 2012 in Uganda and compared the measured GPS-TEC data with the predictions of the IRI-2012 model with NeQuick options. They found that the IRI-2012 NeQuick model did not respond to the effects of the geomagnetic storm.

Figure 6.16 depicts the diurnal variation of the DVTEC during the disturbed periods investigated in this study. In general, no significant differences in the DVTEC values for the quiet and disturbed periods were observed. Similar to those observed

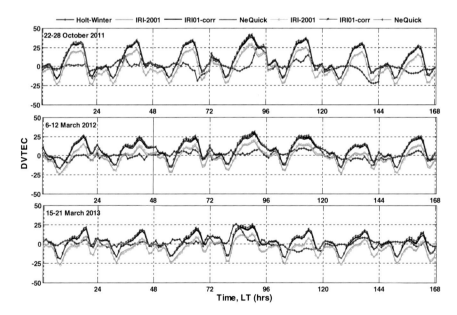

Fig. 6.16 Daily variations of DVTEC during the disturbed period. The IRI-2016 (IRI-2001, IRI01-corr and NeQuick) model values are plotted in solid lines and the IRI-2012 (IRI-2001, IRI01-corr and NeQuick) model values are plotted in dotted lines

for the quiet periods, the DVTEC values for the period of low solar activity in October 2011 were lower than those for 2012 and 2013. The maximum DVTEC values obtained from the modelled VTEC were 27 and 30 (IRI-2001) and 38 and 39 (IRI01-corr).

The measured and modelled VTEC values for the geomagnetic storm event over the UUMK, MUKH and TGPG stations on 18 March 2013 were tested to investigate and evaluate the performances of the IRI-2016, IRI-2012 (IRI-2001, IRI01-corr and NeQuick) and Holt–Winter models. The top panel in Fig. 6.17 corresponds to the diurnal variation of the measured and modelled VTEC values for the UUKM, MUKH and TGPG stations, whereas the two bottom panels correspond to the %Dev and DVTEC values of the corresponding models. The diurnal VTEC trend for the UUMK, MUKH and TGPG stations showed that the GPS-TEC measurements for UUMK were high for the early mornings and post noon. By contrast, the minimum VTEC values were observed for the MUKH station. The IRI01-corr and NeQuick values for IRI-2016, IRI-2012 and the Holt–Winter model were underestimated relative to the GPS-TEC data for the UUMK and TGPG stations in the early morning period between 00:00–04:00 LT. Likewise, for the MUKH station, the Holt–Winter model underestimated the GPS-TEC data, whereas the IRI01-corr and NeQuick models overestimated the GPS-TEC data. The IRI-2001 model showed a slight underestimation for the UUMK station and overestimation for the MUKH and TGPG stations.

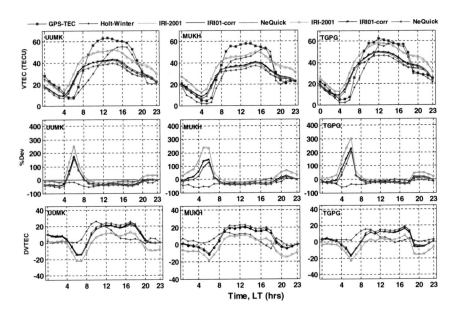

Fig. 6.17 Hourly diurnal variation during the geomagnetic storm on 18 March 2013 (top panel), %Dev (middle panel) and DVTEC (bottom panel) of IRI-2016, IRI-2012 top-side options and the Holt–Winter model for the UUMK, MUKH and TGPG stations. IRI-2016 (IRI-2001, IRI01-corr and NeQuick) model values are plotted in solid lines and IRI-2012 (IRI-2001, IRI01-corr and NeQuick) model values are plotted in dotted lines

For all stations, the IRI top-side options overestimated the GPS-TEC data for 04:00 to 08:00 LT and then underestimated the GPS-TEC data for 08:00 to 18:00 LT. They then overestimated the GPS-TEC data. The Holt–Winter model, however, mostly underestimated the GPS-TEC data. All models showed the maximum underestimation for the UUMK station.

In general, the hourly %Dev trends for the UUMK, MUKH and TGPG stations during the disturbed periods were almost identical. The %Dev in Fig. 6.16 indicates that the IRI top-side options displayed the highest %Dev for the morning hours between 03:00–07:00 LT. The IRI-2001 model attained the highest %Dev values, followed by the IRI01-corr and NeQuick models, which showed similar patterns. Finally, the Holt–Winter model displayed the lowest %Dev. The %Dev for the MUKH station was slightly lower than that for the UUMK and TGPG stations. However, the variation of the %Dev in Fig. 6.15 did not clearly depict the differences between the performances of the IRI-2016 and IRI-2012 top-side options.

The hourly variation of the VTEC values for the geomagnetic storm day for the UUMK, MUKH and TGPG stations modelled with the IRI models indicated that the IRI-2016 (NeQuick) model displayed lower DVTEC and better modelling capability than the AKUM IRI-2012 (NeQuick) model. For the UUMK station, the IRI-2016 (IRI01-corr) showed better agreement with the GPS-TEC data and lower DVTEC than the IRI-2012 (IRI01-corr) model. For the MUKH station, the IRI01-corr model did not show a significant difference from the IRI-2016 and IRI-2012 models, whereas for the TGPG station, IRI-2012 (IRI01-corr) showed a slightly lower DVTEC than IRI-2016 (IRI01-corr). NeQuick had the maximum DVTEC, whereas the IRI01-corr model had the minimum. The highest values of DVTEC were 21 and 21 (IRI-2001), 22 and 23 (IRI01-corr) and 25 and 26 (NeQuick) for the UUMK station. The next-highest DVTEC values were 18 and 17 (IRI-2001), 20 and 20 (IRI01-corr) and 21 and 23 (NeQuick) for the MUKH station. The highest DVTEC values for the TGPG station obtained with the IRI-2016 and IRI-2012 models were 23 and 22 (IRI-2001), 23 and 22 (IRI01-corr) and 18 and 26 (NeQuick). The Holt–Winter model showed the highest DVTEC of 26 for the UUMK station and values of 15 and 25 for the MUKH and TGPG stations, respectively. For all the investigated periods, the IRI-2016 model had a slightly better prediction capability than IRI-2012 during the period of disturbance. The Holt–Winter model showed an increasing trend in the DVTEC values for 08:00–12:00 LT at the UUMK, MUKH and TGPG stations. For the remaining periods, the Holt–Winter model displayed lower DVTEC values than the IRI-2016 and IRI-2012 (IRI-2001, IRI01-corr and NeQuick) top-side options.

6.9 Summary

The statistical Holt–Winter model was used to estimate and model VTEC variations at three different latitudes. The accuracy of the model was evaluated under two conditions. The results illustrated the effectiveness of the Holt–Winter model. The forecasts of the Holt–Winter model showed a good agreement with GPS-TEC measurements for the quiet period and slight underestimation and overestimation for the disturbed period. The maximum MAPE was 4.5% for the quiet period and 8.7% for the disturbed period. These results proved that the Holt–Winter model provided accurate results for quiet and disturbed periods at different locations. In addition, the performance of the Holt–Winter model was compared with that of the IRI-2012 model. The VTEC measured over the Langkawi station in northern Malaysia during 2014 was compared with the VTEC forecasted by the Holt–Winter model and the IRI-2012 top-side options. The overall VTEC reached its minimum values at 5:00–6:00 LT during sunrise and its maximum values at 13:00–17:00 LT shortly afternoon. The IRI-2001 predictions for the early morning until afternoon were overestimated relative to the GPS-TEC values and then gradually decreased until night-time. The NeQuick and IRI01-corr models showed similar trends and underestimated values for most hours except during the afternoon, whereas the trend of the Holt–Winter model results was similar to that of the GPS-TEC measurements. The highest and lowest monthly VTEC values were observed for March and July, respectively. The peak values given by IRI-2001, NeQuick and IRI01-corr were significantly underestimated relative to the GPS-TEC measurements. In general, NeQuick and IRI01-corr underestimated the values for all months. IRI-2001 overestimated the VTEC for June, July and August and underestimated the VTEC for the remaining months. Meanwhile, the Holt–Winter model similar trends and slightly underestimated VTEC values for the year. The GPS-TEC measurements showed seasonal variation and had the lowest values during summer, the next-lowest values during winter and the maximum values during the equinox. The comparative analysis of the NeQuick, IRI01-corr and the Holt–Winter model revealed that the monthly %Dev of the IRI-2001 model for the morning hours of 4:00–8:00 LT was high and that for the months of May, June and July increased significantly, whereas that for the times between 13:00–19:00 LT was less than that of the values forecasted by IRI01-corr and NeQuick models. The maximum value of the seasonal %Dev was observed in the summer. In addition, for the chosen period, the accuracies of the Holt–Winter model, IRI-01-corr, NeQuick, and IRI-2001 were approximately 95%, 75%, 73% and 66%, respectively. Hence, this study proved the effectiveness of the Holt–Winter model in providing good results with high accuracy during diurnal, monthly and seasonal variations. The IRI01-corr and NeQuick models provided the next-best results, whereas the IRI-2001 model showed poor prediction results for Langkawi, Malaysia. The performance of the Holt–Winter model for the quiet and disturbed periods was compared with that of the IRI-2012 model by using the IRI-2001, IRI01-corr and NeQuick top-side options. The variations in ionospheric VTEC over three different stations during geomagnetically quiet and disturbed periods with ascending and high solar activity in October 2011, March 2012 and March 2013 and the improvement in the performance of the most recent IRI model, namely, IRI-2016 (IRI-2001, IRI01-corr and NeQuick), were investigated and compared with those of IRI-2012 by using the NeQuick, IRI-2001 and IRI01-corr top-side electron density options followed by the time series Holt–Winter model. The ionospheric VTEC modelled by IRI-2016 had a slight improvement from that modelled by IRI-2012. Differences were observed for the post noon and night-time, and the VTEC for early morning hours modelled by both IRI models were almost similar. Comparing the prediction capacity of IRI-2016 and IRI-2012 for daily quiet and disturbed periods showed that IRI-2016 predictions showed better agreement with GPS-TEC measurements. The overall results showed that the model's prediction performance for the period of high solar activity in 2013 was better than that for the period of ascending solar activity. Comparison amongst the prediction performances of the IRI-2016, IRI-2012 and the Holt–Winter models for periods of high solar activity revealed that all the results of the models for quiet periods showed better agreement with GPS-TEC measurements than for disturbed periods. The agreement of the time-series Holt–Winter model with GPS-TEC VTEC during the quiet and disturbed periods was the best, followed by that of IRI01-corr, then by that of the NeQuick and IRI-2001 options. This result could help represent the model error of data assimilation, which is useful for improving ionospheric models. In addition, this comparative study is important for identifying and selecting suitable models to improve the accuracy of GPS positioning over equatorial regions.

References

1. A.O. Adewale, E.O. Oyeyemi, J.O. Adeniyi, A.B. Adeloye, O.A. Oladipo, Comparison of total electron content predicted using the IRI-2007 model with GPS observations over Lagos, Nigeria. Indian J. Radio Space Phys. **40**(1), 21–25 (2011). http://nopr.niscair.res.in/handle/123 456789/11194
2. M. Aggarwal, TEC variability near northern EIA crest and comparison with IRI model. Adv. Space Res. **48**(7), 1221–1231 (2011). https://doi.org/10.1016/j.asr.2011.05.037
3. D. Bilitza, D. Altadill, V. Truhlik, V. Shubin, I. Galkin, B. Reinisch, X. Huang, International Reference Ionosphere 2016: from ionospheric climate to real-time weather predictions. Space Weather **15**(2), 418–429 (2017). https://doi.org/10.1002/2016SW001593
4. D. Bilitza, Evaluation of the IRI-2007 model options for the topside electron density. Adv. Space Res. **44**(6), 701–706 (2009). https://doi.org/10.1016/j.asr.2009.04.036
5. D. Bilitza, International reference ionosphere 2000. Radio Sci. **36**(2), 261–275 (2001). https://doi.org/10.1029/2000RS002432
6. D. Bilitza, D. Altadill, Y. Zhang, C. Mertens, V. Truhlik, P. Richards, L.-A. McKinnell, B. Reinisch, The international reference ionosphere 2012—a model of international collaboration. J. Space Weather Space Clim. **4**, A07 (2014). https://doi.org/10.1051/swsc/2014004
7. J.A. Bittencourt, V.G. Pillat, P.R. Fagundes, Y. Sahai, A.A. Pimenta, LION: A dynamic computer model for the low-latitude ionosphere. Ann. Geophys. **25**(11), 2371–2392 (2007)
8. M. Chakraborty, S. Kumar, B.K. De, A. Guha, Latitudinal characteristics of GPS derived ionospheric TEC: a comparative study with IRI 2012 model. Ann. Geophys. **57**(5), 1–13 (2014). https://doi.org/10.4401/ag-6438
9. N.A. Elmunim, M. Abdullah, A.M. Hasbi, S.A. Bahari, *Comparison of GPS TEC Variations with Holt-Winter Method and IRI-2012 Over Langkawi* (Advances in Space Research, Malaysia, 2016). https://doi.org/10.1016/j.asr.2016.07.025
10. J.B. Habarulema, L.A. McKinnell, D. Burešová, Y. Zhang, G. Seemala, C. Ngwira, J. Chum, B. Opperman, A comparative study of TEC response for the African equatorial and mid-latitudes during storm conditions. J. Atmos. Solar Terr. Phys. **102**, 105–114 (2013). https://doi.org/10.1016/j.jastp.2013.05.008
11. S.P. Karia, N.C. Patel, K.N. Pathak, Comparison of GPS based TEC measurements with the IRI-2012 model for the period of low to moderate solar activity (2009–2012) at the crest of equatorial anomaly in Indian region. Adv. Space Res. **55**(8), 1965–1975 (2015). https://doi.org/10.1016/j.asr.2014.10.026

12. S. Kumar, S. Priyadarshi, G.G. Krishna, A.K. Singh, GPS-TEC variations during low solar activity period (2007–2009) at Indian low latitude stations. Astrophys. Space Sci. **339**(1), 165–178 (2012). https://doi.org/10.1007/s10509-011-0973-6

13. S. Kumar, A.K. Singh, J. Lee, Equatorial Ionospheric Anomaly (EIA) and comparison with IRI model during descending phase of solar activity (2005–2009). Adv. Space Res. **53**(5), 724–733 (2014). https://doi.org/10.1016/j.asr.2013.12.019

14. S. Kumar, E. Tan, S. Razul, C.M. See, D. Siingh, Validation of the IRI-2012 model with GPS-based ground observation over a low-latitude Singapore station. Earth Planets Space **66**(1), 17 (2014). https://doi.org/10.1186/1880-5981-66-17

15. M. Limberger, W. Liang, M. Schmidt, D. Dettmering, U. Hugentobler, Regional representation of F2 Chapman parameters based on electron density profiles. Ann. Geophys. **31**(12), 2215–2227 (2013). https://doi.org/10.5194/angeo-31-2215-2013

16. E. Mengistu, M.B. Moldwin, B. Damtie, & M. Nigussie, (2019). The performance of IRI-2016 in the African sector of equatorial ionosphere for different geomagnetic conditions and time scales. J Atmos Sol-Terr Phys, **186**, 116–138

17. N.C. Patel, S.P. Karia, K.N. Pathak, Evaluation of the improvement of IRI-2016 over IRI-2012 at the India low-latitude region during the ascending phase of cycle 24. Adv. Space Res. **63**(6), 1860–1881 (2019). https://doi.org/10.1016/j.asr.2018.10.008

18. G.K. Seemala, C.E. Valladares, Statistics of total electron content depletions observed over the South American continent for the year 2008. Radio Sci. **46**(5) (2011). https://doi.org/10.1029/2011RS004722

19. A.Y. Tariku, Assessment of improvement of the IRI model over Ethiopia for the modeling of the variability of TEC during the period 2013–2016. Adv. Space Res. **63**(5), 1634–1645 (2019). https://doi.org/10.1016/j.asr.2018.11.014

20. Y.A. Tariku, TEC prediction performance of IRI-2012 model during a very low and a high solar activity phase over equatorial regions, Uganda. J. Geophys. Res. A: Space Phys. **120**(7), 5973–5982 (2015). https://doi.org/10.1002/2015JA021203

21. Y.A. Tariku, Patterns of GPS-TEC variation over low-latitude regions (African sector) during the deep solar minimum (2008 to 2009) and solar maximum (2012 to 2013) phases. Earth Planets Space **67**(1), 1 (2015)

22. M.A. Tariq, M. Shah, M. Ulukavak, T. Iqbal, Comparison of TEC from GPS and IRI-2016 model over different regions of Pakistan during 2015–2017. Adv. Space Res. **64**(3), 707–718 (2019). https://doi.org/10.1016/j.asr.2019.05.019

23. K. Venkatesh, P.R. Rao, P. Saranya, Vertical electron density and topside effective scale height (HT) variations over the Indian equatorial and low latitude stations. Ann (2011). https://www.researchgate.net/profile/Drk_Venkatesh/publication/253171941_Vertical_electron_density_and_topside_effective_scale_height_(HT)_variations_over_the_Indian_equatorial_and_low_latitude_stations/links/53e3d43f0cf2fb74870dcca5.pdf

24. M.L. Zhang, S.M. Radicella, J.K. Shi, X. Wang, S.Z. Wu, Comparison among IRI, GPS-IGS and ionogram-derived total electron contents. Adv. Space Res. **37**(5), 972–977 (2006). https://doi.org/10.1016/j.asr.2005.01.113

Printed in the United States
by Baker & Taylor Publisher Services